INFRARED RECEPTORS AND THE TRIGEMINAL SENSORY SYSTEM

INFRARED RECEPTORS AND THE TRIGEMINAL SENSORY SYSTEM

Edited by

Shin-ichi Terashima, MD, Ph.D.
University of the Ryukyus School of Medicine
Okinawa-ken, Japan

and

Richard C. Goris, Ph.D.
Yokohama City University School of Medicine
Yokohama, Japan

 harwood academic publishers
Australia • Canada • China • France • Germany • India • Japan • Luxembourg
Malaysia • The Netherlands • Russia • Singapore • Switzerland • Thailand

Amsteldijk 166
1st Floor
1079 LH Amsterdam
The Netherlands

British Library Cataloguing in Publication Data

Infrared receptors and the trigeminal sensory system
  1. Trigeminal nerve 2. Neural receptors
  I. Terashima, Shin-ichi   II. Goris, Richard C.
    612.8'19

  ISBN   90-5702-217-6

Front cover photograph:

The sensory membrane of a snake pit organ

CONTENTS

FOREWORD

Professor S. Terashima and Dr R.C. Goris have accomplished great strides in investigating infrared sensor systems, employing physiological methodologies, over the last forty years. In 1987, Professor Terashima published a book entitled *Infrared Sensory System* (University of the Ryukyus), which compiled his major works between 1968 and 1984.

The collective works presented in this book comprise nineteen publications by Professor Terashima and his coworkers from the past ten years. We are extremely pleased to support this publication, in order to provide the readership with a single compilation of his more recent works, which have previously been published in a number of different periodicals. We believe that this science and the subject area are of interest not only to the mainstream neurophysiology community, but many biochemists, biophysicists and microelectronic engineers, and that promulgation of this research will better facilitate an understanding of biophysiological transmission systems.

We, the Asian Office of Aerospace Research and Development (AOARD), a detachment of the USAF Office of Scientific Research (AFOSR), are delighted to support this publication, as one of our roles in assisting interdisciplinary interaction among scientists and promoting the advancement of basic science. Established in 1992, the AOARD in Tokyo has been supporting the Asian science community by organizing international symposia and workshops in the materials sciences, biorobotics, microelectronics and many other advanced technology areas, and has published many of these proceedings in internationally renowned periodicals. We additionally support visits by Asian scientists to USAF and other US governmental laboratories, as well as offer grants and R&D contracts in basic sciences for the promotion of collaborative research.

In publishing this book, Gordon and Breach Science Publishers have been very helpful and cooperative with us. We sincerely express our thanks to them for accepting this publication.

Shiro Fujishiro
AFOSR/AOARD

PREFACE

The goal of this monograph is to provide an overview of recent developments in our studies on snake infrared receptors and their associated nervous systems. We are attempting to throw light on how the central and peripheral nervous systems work to process the sensory information of infrared images. We hope that the book will be useful to both physiologists and anatomists interested in somatosensory systems.

Our publications make up more than half of the works published around the world on reptilian infrared reception. Our work has been done entirely in Japan, partly with the help of foreign students, to whom we are grateful, and mainly with snakes endemic to Japan and adjacent countries. The research covers the fields of physiology, anatomy, and histochemistry, so that the papers are necessarily spread around among many specialist journals. The convenience of having such papers gathered together in a single volume has been well borne out in our first book, *Infrared Sensory System*, published by the University of the Ryukyus. It is our belief that lecturers on infrared receptors will find here the information they need, while graduate students and others interested in infrared reception will be able to get a sound integrated account of results of our research in this area.

We wish to thank the Asian Office of Aerospace Research and Development for their support, and the following publishers and journals for permission to use copyright material: *The Anatomical Record, Cellular and Molecular Neurobiology, Brain Research, Animal Eye Research, Somatosensory and Motor Research, Neuroscience Research, The Journal of Comparative Neurology, Neuroscience Letters, Biomedical Research, Japanese Journal of Physiology*, The Herpetological Society of Japan, *Cell and Tissue Research, Neuroscience*, and *Zeitschrift für Hirnforschung*.

Shin-ichi Terashima
Richard C. Goris

ACKNOWLEDGEMENTS

Chapter 1:
Reprinted from *The Anatomical Record*, Vol. 246, pp. 135–146, F. Amemiya *et al.*, Copyright © 1996, with kind permission from Wiley-Liss, Inc., a division of John Wiley & Sons, Inc.

Chapter 2:
Reprinted from *Cellular and Molecular Neurobiology*, Vol. 17, pp. 195–206, S. Terashima and A.-Q. Zhu, Copyright © 1997, with kind permission from Plenum Publishing Corporation.

Chapter 3:
Reprinted from *Brain Research*, Vol. 713, pp. 168–177, P.-J. Jiang and S. Terashima, Copyright © 1996, with kind permission from Elsevier Science-NL, Sara Burgerhartstraat 25, 1055 KV Amsterdam, The Netherlands.

Chapter 4:
Reprinted from *Animal Eye Research*, Vol. 15, pp. 13–25, F. Amemiya, R.C. Goris, *et al.*, Copyright © 1996, with kind permission from the Japanese Society of Comparative Ophthalmology.

Chapter 5:
Reprinted from *Somatosensory and Motor Research*, Vol. 12, pp. 299–307, P.-J. Jiang and S. Terashima, Copyright © 1995, with kind permission from The Guilford Press.

Chapter 6:
Reprinted from *Somatosensory and Motor Research*, Vol. 12, pp. 143–150, S. Terashima, P.-J. Jiang *et al.*, Copyright © 1995, with kind permission from The Guilford Press.

Chapter 7:
Reprinted from *Neuroscience Research*, Vol. 22, pp. 287–295, M. Sekitani-Kumagai, T. Kadota, R.C. Goris, *et al.*, Copyright © 1995, with kind permission from Elsevier Science Ireland Ltd.

Chapter 8:
Reprinted from *The Journal of Comparative Neurology*, Vol. 360, pp. 621–633, Y.-F. Liang, S. Terashima and A.-Q. Zhu, Copyright © 1995, with kind permission from Wiley-Liss, Inc., a subsidiary of Wiley & Sons, Inc.

Chapter 9:
Reprinted from *Neuroscience Research*, Vol. 22, pp. 315–323, S. Kobayashi, F. Amemiya, *et al.*, Copyright © 1995, with kind permission from Elsevier Science Ireland Ltd.

Chapter 10:
Reprinted from *Biomedical Research*, Vol. 16, pp. 411–421, F. Amemiya, R.C. Goris, Y. Matsuda, *et al.*, Copyright © 1995, with kind permission from the Biomedical Research Foundation.

Chapter 11:
Reprinted from *Neuroscience Letters*, Vol. 179, pp. 33–36, S. Terashima and Y.-F. Liang, Copyright © 1994, with kind permission from Elsevier Science Ireland Ltd.

Chapter 12:
Reprinted from *Somatosensory and Motor Research*, Vol. 11, pp. 169–181, S. Terashima and Y.-F. Liang, Copyright © 1994, with kind permission from The Guilford Press.

Chapter 13:
Reprinted from the *Japanese Journal of Physiology*, Vol. 43, Suppl. I, pp. 267–274, S. Terashima and Y.-F. Liang, Copyright © 1993, with kind permission from the Center for Academic Publications Japan.

Chapter 14:
Reprinted from *The Journal of Comparative Neurology*, Vol. 328, pp. 88–102, Y.-F. Liang and S. Terashima, Copyright © 1993, with kind permission from Wiley-Liss, Inc., a subsidiary of Wiley & Sons, Inc.

Chapter 15:
Reprinted from *Brain Research*, Vol. 597, pp. 350–352, S. Kobayashi, R. Kishida, *et al.*, Copyright © 1992, with kind permission from Elsevier Science - NL, Sara Burgerhartstraat 25, 1055 KV Amsterdam, The Netherlands.

Chapter 16:
Reprinted from *Current Herpetology in East Asia*, pp. 8–16, R.C. Goris, T. Kadota and R. Kishida, Copyright © 1989, with kind permission from The Herpetological Society of Japan.

Chapter 17:
Reprinted from *Cell and Tissue Research*, Vol. 253, pp. 311–317, T. Kadota, R. Kishida, R.C. Goris and T. Kusunoki, Copyright © 1988, with kind permission from Springer-Verlag GmbH & Co. KG.

Chapter 18:
Reprinted from *Neuroscience*, Vol. 23, No. 2, pp. 685–691, S.Terashima, Copyright © 1987, with kind permission from Pergamon Journals Ltd.

Chapter 19:
Reprinted from *Journal für Hirnforschung*, Vol. 28, pp. 27–43, T. Kusunoki, R. Kishida, *et al.*, Copyright © 1987, with kind permission from Akademie Verlag GmbH.

Ultrastructure of the Crotaline Snake Infrared Pit Receptors: SEM Confirmation of TEM Findings

FUMIAKI AMEMIYA, TATSUO USHIKI, RICHARD C. GORIS, YOSHITOSHI ATOBE, AND TOYOKAZU KUSUNOKI

Department of Anatomy, Yokohama City University School of Medicine, Yokohama-shi, Japan (F.A., R.C.G., Y.A., T.K.), and Department of Anatomy, Niigata University School of Medicine, Niigata-shi, Japan (T.U.)

ABSTRACT   **Background:** Crotaline snakes possess a pair of infrared-sensing pit organs that aid the eyes in the detection and apprehension of prey. The morphology of the receptors in the pit organs has been studied by light and transmission electron microscopy, and the ultrastructure of the receptors has been inferred from the results of this work. But this theoretical reconstruction has never been confirmed by any kind of three-dimensional imaging.

Methods: We treated the receptor-containing membrane of the pit organs with potassium hydroxide to remove collagen and expose the receptors, which we then viewed by scanning electron microscopy.

Results: We were able to obtain three-dimensional views of all structures previously reported to exist within the receptor-containing membrane: terminal nerve masses formed from free nerve endings, supporting Schwann cells within the nerve masses, unmyelinated and myelinated nerve fibers, a capillary bed, and vacuole cells.

Conclusions: By providing the first three-dimensional views of the infrared receptors, we have confirmed that previous theoretical reconstructions of the receptors were substantially correct and have provided new evidence of the spatial arrangement of the receptors in a monolayer array. © 1996 Wiley-Liss, Inc.

Key words: Crotaline snakes, Infrared reception, Pit receptors, TEM, SEM

It is notoriously difficult to construct a three-dimensional image of ultramicroscopic structures simply on the basis of transmission electron microscopic (TEM) ultrathin sections. When such structures exist on the surface of some organ or organism, the scanning electron microscope (SEM) can provide three-dimensional data that, combined with the TEM data, provide a valuable aid for such reconstruction. However, the difficulty is compounded when the target structures exist beneath the surface or deep within an organ.

Snakes of the subfamily Crotalinae of the family Viperidae possess a pair of infrared sensor organs that aid the eyes in detecting and capturing prey (Fig. 1; for a review, see Molenaar, 1992). The receptors are contained in a thin membrane suspended within the pit. Several attempts have been made to provide an idea of the physical form of these receptors, most notably those of Bullock and Fox (1957) and Terashima et al. (1970). Terashima et al. used TEM photographs to provide a plausible reconstruction of the receptors, which they called "terminal nerve masses" (TNMs). This reconstruction has served as the basis for interpretation of the results of practically all work on the pit organs in the succeeding 25 years.

In the present work we have used advanced SEM techniques (Ushiki and Murakumo, 1991) that permit direct, three-dimensional observation of the TNMs, confirming not only the conjectural reconstruction of Terashima et al. (1970) but also subsequent work on the ultrastructure of these organs and related structures and on the infrared receptor organs in boid snakes (Amemiya et al., 1996).

MATERIALS AND METHODS

We used four specimens of *Agkistrodon blomhoffii*, a small pit viper indigenous to Japan, Korea, and eastern China. Two were used for TEM work and the other two for SEM work. All specimens were anesthetized with halothane and perfused through the right aortic arch, first with physiological saline solution, then with fixative containing 2% paraformaldehyde and 2.5% glutaraldehyde in 0.1 M phosphate buffer at pH 7.4. The receptor-containing pit membranes were then dissected out together with the scales of the pit opening

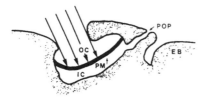

Fig. 1. Schematic representation of a cross section through the infrared receptor organ (pit organ) of a crotaline snake. Infrared rays enter in the direction shown by the long arrows and impinge on the receptors contained in the thin membrane suspended between the inner and outer chambers.

and immersed in the same fixative, 1 day for TEM specimens and several days for SEM specimens.

The scales were removed from the TEM specimens, and the pit membranes alone were immersed in 2% osmium tetroxide for 1 hr, dehydrated in ethanol, and embedded in a mixture of Epon and Araldite or in Luft's Epon mixture. One-micron semithin sections stained with toluidine blue were used for histologic examination and for selecting typical areas for ultrathin sections. The ultrathin sections were then stained for electron microscopy with uranyl acetate and lead citrate.

The SEM specimens were treated to remove collagen by immersion in potassium hydroxide at 60–65°C for 8–10 min (Ushiki and Murakumo, 1991). They were then washed in the phosphate buffer for at least 1 hr, immersed in an aqueous solution of tannic acid for 2–3 hr, rinsed in distilled water for 1 hr, and finally immersed in a 1% aqueous solution of osmium tetroxide at room temperature for 3 hr (Murakami, 1973). Next, the specimens were dehydrated in ethanol, transferred to isoamyl acetate, and critical-point dried in liquid carbon dioxide. Under a dissecting microscope, the pit membrane was cut away from its supporting scales, stripped of its epithelium on the inner and outer surfaces, and mounted on metal stubs with double-sided adhesive tape. The tissue was then sputter-coated with platinum-palladium alloy and observed with a scanning electron microscope.

Abbreviations

cp	capillary
EB	eyeball
fb	fibroblast
IC	inner chamber
nf	nerve fiber
mf	myelinated nerve fiber
OC	outer chamber
OCL	outer cornified layer
OCT	outer connective tissue layer
OEL	outer epithelial layer
PM	pit membrane
POP	preocular pore
Sch	Schwann cell
TNM	terminal nerve mass
uf	unmyelinated nerve fiber
VC	vacuole cell

RESULTS

Semithin Sections

Light microscopy of semithin sections showed the TNMs arranged in a single layer directly beneath the outer epithelium of the pit membrane (Fig 2, arrows). Directly below the TNMs and sometimes between them are a large number of capillary blood vessels (cp). In the midst of the TNMs, Schwann cell nuclei (Sch) can be discerned. Farthest from the outer epithelium, below the TNMs and capillary network, are groups of myelinated nerve fibers (mf) that presumably branch to innervate the TNMs. The empty space between the receptor layer and the inner epithelium is an artifact of the sectioning.

TEM Sections

In a typical horizontal section through the pit membrane (Fig. 3), the following structures could be identified from top to bottom: outer cornified layer (OCL), outer epithelial layer (OEL), outer connective tissue layer (OCT), terminal nerve mass layer (TNM), a capillary (cp), and a group of myelinated fibers (mf). A TNM can be seen below the OCT at the center of the figure, with its individual terminals clearly delimited by connective tissue and enveloped here and there with Schwann cell cytoplasm (arrows) containing many dense bodies, as described by Terashima et al. (1970). In addition, an inner connective tissue layer, an inner epithelial layer, and an inner cornified layer are visible (Hirosawa, 1980), but these are not shown here.

In Figure 4, which is another section of the same material, the basement membrane (arrowheads) delimiting the OEL from the TNM layer is very evident. This finding is characteristic of crotaline snakes, in contrast with boids, where the TNMs are located above the basement membrane (Amemiya et al., 1996). Within the OCT are a number of specialized vacuole cells (VC), which are present in Figure 3, but in Figure 4 they can be seen in the process of exocytosis (arrows), presumably secreting the intercellular liquid marked by asterisks. The TNM at the center of the figure is a very typical one, measuring about 20 μm in long diameter and showing the characteristic dense packing of mitochondria. The mitochondria themselves show several different configurations: discharged (a), partially discharged or partially charged (b), and charged (c). Below the TNM appears part of a capillary (cp), and below that a myelinated nerve fiber (mf).

SEM sections

In horizontal cross section, the monolayer array of TNMs extrapolated from TEM sections is clearly visible (Fig. 5). The TNMs are covered by the outer connective tissue (OCT), and even vacuole cells (VC) in the process of exocytosis (arrowhead) are present in the picture. Myelinated fibers (mf), which will eventually branch to form the TNMs, course beneath the array. A more highly magnified view of a TNM (Fig. 6) shows the typical waferlike form so suitable for functioning in a sensory array and the strongly segmented structure apparent from Figures 3 and 4. This TNM measures about 40 μm from left edge to right edge.

In a view from the inner chamber side (Figs. 7, 8), the myelinated fibers course beneath the TNMs and then

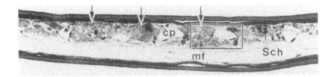

Fig. 2. Semithin section through the pit membrane, stained with toluidine blue. Arrows point to the terminal nerve masses. The box shows a histological picture similar to that appearing in the SEM photographs of Figures 9 and 10. Bar = 50 μm.

lose their myelin sheath (e.g., at the arrowhead) as they bend upward to form the array of sensors. TEM photographs indicate that Schwann cells insinuate their cytoplasm in between the TNM segments (Terashima et al., 1970), and this finding is corroborated by the magnified view of Figure 8 (Sch and arrowheads). In Figure 8 a fibroblast (fb) appears to cover the point where the myelinated fiber (mf) loses its sheath to branch upward as an unmyelinated fiber (uf).

The intimate apposition of blood capillaries with TNMs and their feeder nerve fibers can be inferred from the boxed area in Figure 2, and we were able to confirm this with clear views from below (i.e., the inner chamber side, Fig. 9) and above (the outer chamber side, Fig. 10) the sensory array. The capillaries (cp) can be clearly identified by the pericytes on their surface (arrowheads), and they form a dense network below and between the TNMs and are virtually in contact with these receptors, suggesting that they have a very important role in the receptor function. The network concept is especially apparent in Figure 9, which shows a junction from which four capillary vessels branch off. The view in Figure 10 is through a hole opened in the outer surface of the pit membrane and shows two capillaries. The one at left is in a rather deep position and is possibly a feeder vessel for the receptor contact vessels such as the one at the right edge of the figure.

DISCUSSION

The SEM photographs presented here provide the first published evidence of the three-dimensional morphology of the fundamental structures making up the infrared receptor system in pit vipers. In brief, the nerves innervating the pit membrane enter as myelinated fibers in the innermost (i.e., closest to the inner chamber) layer of the membrane. These fibers branch repeatedly (Bullock and Fox, 1957; Terashima et al., 1970) and bend toward the outer surface of the membrane. In the process of branching and bending, the fibers lose their myelin coating and terminate in expanded processes immediately beneath the outer epithelial and cornified layers. The terminal process are tightly grouped into discrete bundles that have been aptly named "terminal nerve masses" (Terashima et al., 1970).

Specialized Schwann cells are associated with the TNMs. These are specialized in the sense that instead of producing a myelin sheath they extend their cytoplasmic processes into the interstices between the terminal processes and around the TNMs. The function of the Schwann cell cytoplasm extensions may be to provide support to hold the TNMs and their individual components in place and to maintain a microenvironment suitable for axon terminals (Terashima et al., 1970).

Running between and immediately beneath the TNMs are a large number of capillary blood vessels. These vessels supply the high energy requirements of the TNMs and act as a cooling network, as described by Amemiya et al. (1996).

The SEM observations are in excellent accord with the TEM and light micrographs we present here. They also are excellent confirmation of the observations, terminology, and reconstructions of previous workers. Bullock and Fox (1957) published the first histological study of the receptors and described them with the term "palmate structures." Figure 7 shows that this was indeed an appropriate description, in spite of the fact that their silver staining did not extend all the way to the end of the terminals, as pointed out by Bleichmar and de Robertis (1962). Terashima et al. (1970) published what is now a classic reconstruction of the TNMs based on electron micrographs, and the accuracy of this reconstruction is also borne out by our Figure 7.

Of the functional structures described by other workers, only two do not appear in our SEM photographs: the mitochondria packed inside the TNMs and the vacuoles of vacuole-containing cells (Bleichmar and de Robertis, 1962; Hirosawa, 1980).

Mitochondria, of course, cannot be seen in our SEM photographs because we are viewing the outside structure of the TNMs. Osmium maceration techniques (Tanaka and Naguro, 1981) will be necessary to visualize the structure, location, and configuration of the mitochondria by SEM. However, our TEM photographs (Figs. 3, 4) clearly show three stages of activity of the mitochondria: completely discharged, partially discharged (or partially charged), and charged. These three states were noted by Meszler (1970), in a short communication, but this paper is only the second report about these mitochondrial stages. Further work at higher magnification may provide a way of distinguishing the partially discharged state from the partially charged state.

The vacuole cells (VCs) are clearly seen in our Figures 3–5. Bleichmar and de Robertis (1962) commented that these cells may have some accessory, although not essential, role in the function of the

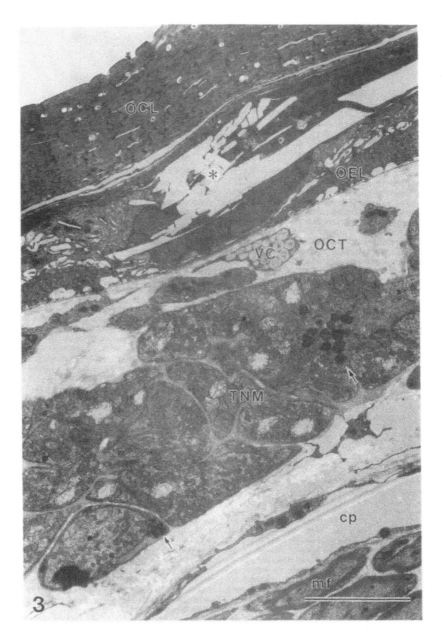

Fig. 3. A horizontal section through the pit membrane showing additional details of the structures identified in Figure 2. An asterisk marks the break in the outer epithelium where shedding normally occurs. Arrows point to Schwann cell cytoplasm containing dense bodies. The black blotch at lower left is an artifact. Bar = 5 μm.

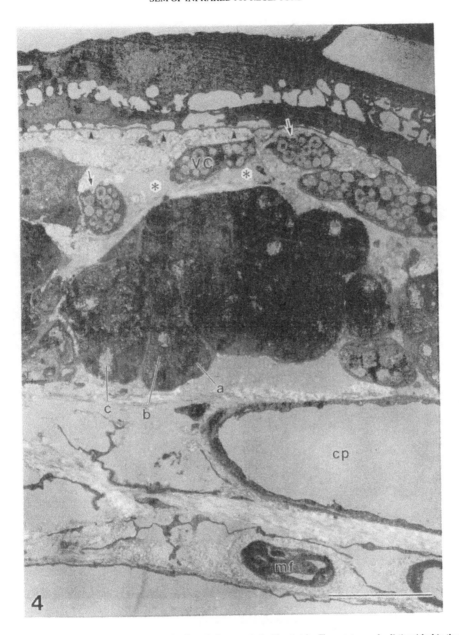

Fig. 4. A section adjacent to that of Figure 3. Arrowheads mark the outer basement membrane. Vacuole cells can be seen in the act of exocytosis (arrows). They may be secreting the interstitial fluid marked with asterisks. Three states can be distinguished in the mitochondria: discharged (a), partially discharged or partially charged (b), and charged (c). Bar = 5 μm.

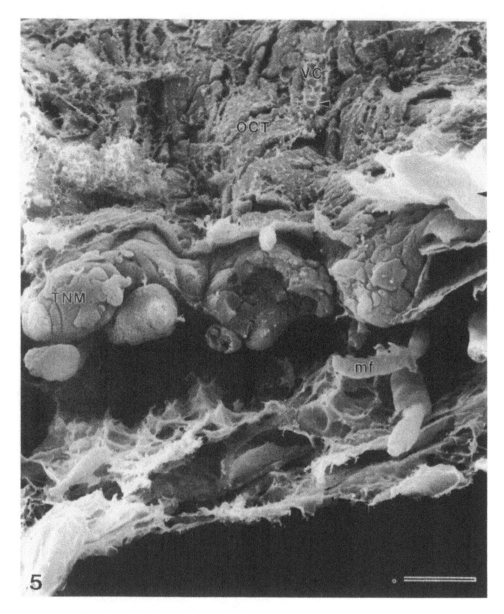

Fig. 5. An SEM cross section through the pit membrane. The arrowhead points to an exocytotic crater in a vacuole cell. Bar = 10 μm.

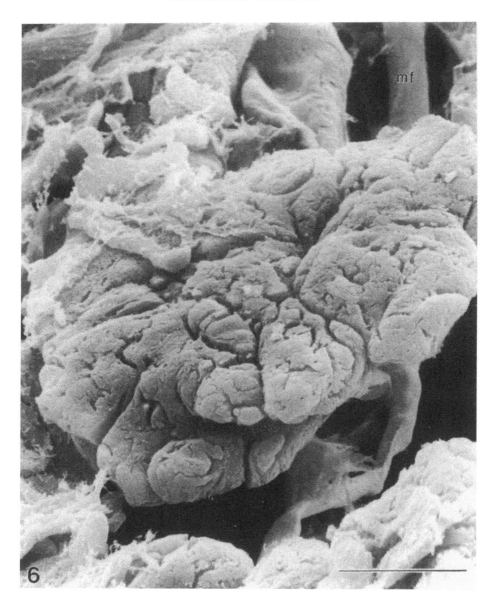

Fig. 6. Close-up of the outer surface of a terminal nerve mass. Bar = 10 μm.

infrared organ. That VCs are functional in some way is clear from the exocytotic picture seen in Figure 4 and 5. Because VCs are found directly in the path traveled by infrared radiation to the receptors, they may be secreting some liquid, (perhaps a lipid) that aids the optic coupling between the cornified epithelium and the re-

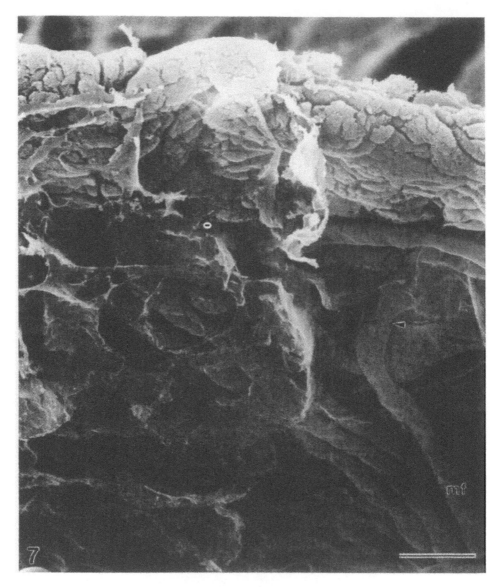

Fig. 7. A terminal nerve mass viewed from beneath and to one side. The arrowhead indicates the point where the myelinated fiber loses its sheath. Bar = 10 μm.

ceptors, like the balsam used to mount specimens in light microscopy.

The presence of Schwann cell cytoplasm enveloping the TNMs and insinuated between the individual components of the TNMs has been noted by all previous electron microscope reports on the pit organs (Bleich-

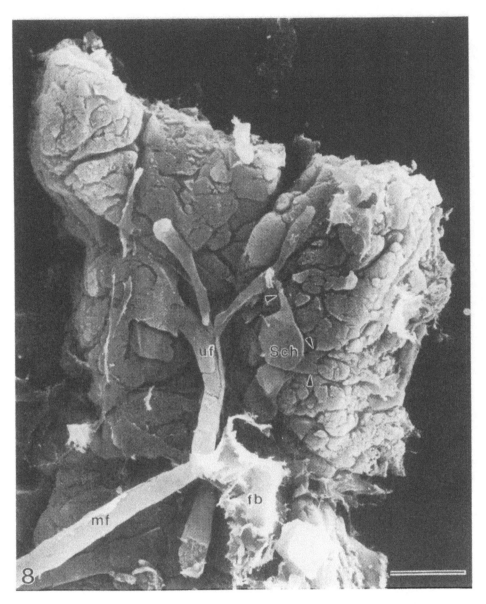

Fig. 8. A terminal nerve mass viewed from beneath. Arrowheads show where the Schwann cell inserts its cytoplasm into the spaces between the segments of the nerve mass. Bar = 10 μm.

F. AMEMIYA *et al.*

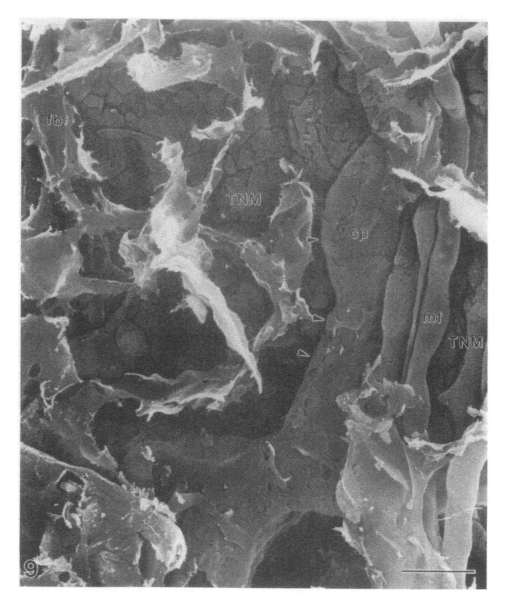

Fig. 9. Another view of terminal nerve masses from beneath, showing the close proximity of the capillary network to the terminal nerve masses. Arrowheads indicate the pericytes of the capillaries. Bar = 10 μm.

Fig. 10. An overall view of the terminal nerve mass array seen from the outer chamber. Bar = 10 μm.

mar and de Robertis, 1962; Terashima et al., 1970; Hirosawa, 1980). This finding is in contrast with the condition found in boid infrared receptors, which are in the outer epithelium and supported by epithelial cells (von Düring, 1974; Amemiya et al., 1996) The reason for the difference may be found in the fact that in snake infrared receptors we have a good case of parallel evolution.

Boids are a primitive type of snake, retaining many characteristics of their lizardlike ancestors, the most prominent of which are the possession of two lungs and the retention of vestiges of the pelvic girdle. Crotalines represent a highly evolved type of snake, with only one (the right) lung remaining, no pelvic girdle, and a highly developed poison apparatus.

In both types of snakes, the infrared reception apparatus is essentially identical: mitochondria-packed terminal nerve masses supplied with energy and cooled by a closely associated capillary network. The difference is found in the location of the TNMs. In boids they are contained in the outer epithelium of the labial scales, where they are supported by the epithelial cells. In addition, they are renewed and shed periodically with every molting cycle (Amemiya et al., 1996).

In contrast, in the crotalines the TNMs are contained in a thin membrane in the loreal region, where they are beneath the basement membrane, and there is no evidence that they are periodically renewed because only the outer generation of the epithelium above the basement membrane is shed at each molting (Landmann, 1986).

That TNMs are suspended in a membrane separated from the rest of the head tissues is also an advantage over the primitive boids because the heat capacity of the receptors is greatly reduced. In other words, heat loss by conduction to the other head tissues is reduced, so that an equal amount of infrared radiation will produce a stronger stimulus in the crotalines than in the boids, making the crotaline infrared organs much more sensitive than their counterparts in the boids.

LITERATURE CITED

Amemiya, F., R.C. Goris, Y. Atobe, N. Ishii, and T. Kusunoki 1996 The ultrastructure of infrared receptors in a boid snake, *Python regius:* Evidence for periodic regeneration of the terminals. Animal Eye Res. *15:*13–25.

Bleichmar, H., and E. de Robertis 1962 Submicroscopic morphology of the infrared receptor of pit vipers. Z. Zellforsch. *56:*748–761.

Bullock, T.H., and W. Fox 1957 The anatomy of the infra-red sense organ in the facial pit of pit vipers. Quart. J. Microsc. Sci. *98(part 2):*219–234.

Hirosawa, K. 1980 Electron microscopic observations on the pit organ of a crotaline snake *Trimeresurus flavoviridis.* Arch. Histol. Jap., *43:*65–77.

Landmann, L. 1986 Epidermis and dermis. In: Biology of the Integument vol. 2: Vertebrates, Part IV: The Skin of Reptiles. J. Bereiter-Hahn, A.G. Matoltsy, and K. Sylvia Richards, eds. Springer-Verlag, Berlin, pp. 150–187.

Meszler, R.M. 1970 Correlation of ultrastructure and function. In: Biology of the Reptilia, vol. 2. C. Gans and T.S. Pearsons, eds. Academic Press, London, pp. 305–314.

Molenaar, G.J. 1992 Sensorimotor integration. In: Biology of the Reptilia, vol. 17: Neurology C. C. Gans and P.S. Ulinski, eds. University of Chicago Press, Chicago, pp. 367–453.

Murakami, T. 1973 A metal impregnation method of biological specimens for scanning electron microscopy. Arch. Histol. Jap., *35:* 323–326.

Tanaka, K., and T. Naguro 1981 High resolution scanning electron microscopy of cell organelles by a new preparation method. Biomed. Res. *2(Suppl.):*63–70.

Terashima, S., R.C. Goris, and Y. Katsuki 1970 Structure of warm fiber terminals in pit membrane of pit vipers. J. Ultrastruct. Res., *31:*494–506.

Ushiki, T., and M. Murakumo 1991 Scanning microscopic studies of tissue elastin components exposed by a KOH-collagenase or simple KOH digestion method. Arch. Histol. Cytol., *54:*427–436.

von Düring, M. 1974 The radiant heat receptor and other tissue receptors in the scales of the upper jaw of boa constrictor. Z. Anat. Entwickl. Gesch., *145:*299–319.

Single Versus Repetitive Spiking to the Current Stimulus of A-β Mechanosensitive Neurons in the Crotaline Snake Trigeminal Ganglion

Shin-ichi Terashima[1,2] and Ai-Qing Zhu

KEY WORDS: sensory neuron; electrophysiological properties; intrasomal recordings.

SUMMARY

1. Intrasomal recordings of potentials produced by current stimulation *in vivo* were made from 24 (A-β) touch and 19 vibrotactile neurons in the trigeminal ganglion of 29 crotaline snakes, *Trimeresurus flavoviridis*.

2. Usually touch neurons responded with a single action potential at the beginning of a prolonged depolarizing pulse, whereas all vibrotactile neurons responded with multiple spikes.

3. The electrophysiological parameters examined were membrane potential, threshold current, input resistance and capacitance, time constant, rebound latency, and its threshold current. Touch neurons had higher input resistance (and lower input capacitance) than vibrotactile neurons.

4. In conclusion, current injection, which elicits a single or multiple spiking, seems a useful way to separate touch neurons from vibrotactile neurons without confirming the receptor response, and some membrane properties are also specific to the sensory modality.

INTRODUCTION

The action potential (AP) forms of the dorsal root ganglion (DRG) neurons of the rat and the cat are related to conduction velocity (CV) [(Harper and Lawson

---

[1] Department of Physiology, University of the Ryukyus School of Medicine, Nishihara-cho, Okinawa 903-01 Japan.
[2] To whom correspondence should be addressed.

13

1985) rat; (Waddell and Lawson, 1990) rat; (Cameron *et al.*, 1986) cat], but there is no good correlation because of overlapping. The AP type of DRG neurons of the cat was related with receptor type by Mendell and his collaborators (Koerber *et al.*, 1989; Rose *et al.*, 1986; Traub and Mendell 1988; Ritter and Mendell, 1992), but this method is not easy because it requires *in vivo* experiments. Ionic channel difference was related to the characteristic form of the AP in DRG neurons and trigeminal root ganglion (TRG) neurons [(Yoshida *et al.*, 1978) mouse; (Waddell and Lawson 1990) rat; (Puil and Spigelman, 1988) guinea pig], but the sensory modality was not clear. If the membrane properties are specific to the sensory modality, and the cell's sensory modality could be identified by observing some sign easy to find in *in vitro* experiments, it would contribute greatly to efficiency.

Martin and Wickelgren (1971) confirmed that by prolonged depolarizing current, all modality-identified neurons elicited one or several APs which were critical enough to be utilized for classification of the sensory neurons in the lamprey. Waddell and Lawson (1990, rat) were unable to classify A neurons into CV groups, which have a short duration of afterhyperpolarization (AHP), by response to depolarizing current. It appears that using this intrasomal depolarizing current sensory neurons can be separated into sensory modality groups, but not into CV groups.

Recently, TRG neurons have been investigated in a few laboratories using intrasomal recording techniques in mammals (Fukuda and Kameyama, 1980; Puil *et al.*, 1986, 1988), but detailed information about the relationship between electrophisiological properties and the sensory modality is still scarse.

Since many types of active membrane properties of identified TRG neurons in snakes could be differentiated, using natural stimulus modalities, into temperature neurons (Terashima and Liang 1991), touch (M) and vibrotactile (V + M) neurons (Terashima and Liang, 1994a), A-δ mechanical nociceptive (mN) neurons (Liang and Terashima 1993), and C mN neurons (Terashima and Liang, 1994b), we wondered if these functional types also differ in their passive membrane properties [time constant (τ), input resistance (R), membrane capacitance (C), and threshold current (T)] in addition to other electrophysiological properties (Terashima and Liang, 1993).

M and V + M neurons, both of which we used in the present experiments, have similar narrow APs with fast CV and short AHP duration of half-decay, and their axons also are similar in morphology (Terashima and Liang, 1994a; Liang *et al.*, 1995). In the present work we concentrated on finding differences in the firing patterns elicited by injected current. If the difference in mean values is positive without overlap, these values could be utilized as an exact and convenient way to distinguish the two kinds of neurons and could be the foundation for further research on membrane channel properties *in vitro*.

A preliminary report has been published elsewhere (Terashima *et al.*, 1992).

METHODS

Animal Preparation

Twenty-nine crotaline snakes, *Trimeresurus flavoviridis*, of both sexes weighing 200–500 g were first anesthetized with halothane and then immobolized with pancuronium (2 mg/kg, im). Artificial respiration with a unidirectional airflow was applied with an aquarium air pump and the flow through the respiratory system was maintained at 0.5 L/min. A few hypodermic needles inserted into the air sac of the lung served as an outlet for the air. Prior to entering the system the air was moistened by passing it through a water bottle. This unidirectional type of artificial respiration avoids respiratory movements, which are often disadvantageous in electrophysiological studies. The animals were fixed with two pairs of snake head holders to provide a stable condition for intracellular recording. A hole was drilled in the lateral skull to expose the trigeminal ganglion through the inner ear. The heart rate was monitored at 50–90 beats/min and the room temperature was kept at 24–26°C, which is the best temperature for this animal (Tanaka *et al.,* 1967). Supplementary doses of pancuronium were administered when necessary during the experiment.

Electrophysiological Recordings

Glass microelectrodes were pulled from thin-walled glass capillaries (o.d. = 1.2 mm, i.d. = 0.94 mm) and filled with $3 M$ KCl solution. The electrode had a resistance of 20–50 MΩ. The electrode was driven by a mechanical micromanipulator and recordings were obtained by a unity-gain high-input impedance preamplifier (Nihon Kohden, MEZ-8201), which senses the intracellular voltage and which is equipped with a bridge circuit for current injection through the recording electrode. Impalement of a neuron, presumably its soma, in the region of the ganglion was accomplished using the microelectrode. Neurons which initially had E_m's more negative than -50 mV and which were able to generate APs of at least 60-mV amplitude in response to injections of depolarizing current pulses were selected for study. The data were displayed and photographed from memory oscilloscopes (Nihon Kohden, VC-11) and stored on an FM data recorder (Sony, A-47; DC-5 kHz). Membrane properties and the AP shapes were assessed using current injected into the cell soma.

Procedure of Detecting a Neuron's Receptive Field and Its Modality

A sudden large and stable DC shift was considered as a first signal that the microelectrode had penetrated a soma. Then the neuron was injected with a positive current pulse to elicit APs or with negative current pulses for recording and measuring the membrane properties. After these procedures, the modality and receptive field were determined by a series of stimulations. Since only infrared-sensitive neurons in the trigeminal ganglion have background discharge, if a neuron with no background discharges was recorded, we used a set of von Frey hairs to stimulate the skin or oral mucosa mechanically. We could identify both M and V + M neurons responding to a threshold of 5- or 10-mg pressure. The former exhibited a train of discharges to a sustained mechanical stimulus of higher intensity, and the latter showed only one or sometimes two discharges at the initial application. A handheld vibrator was used to identify V + M receptive fields. Vibrating stimulation ranged from 1 to 500 Hz was applied to the receptive field, and V + M neurons responded one to one very regularly from 5 up to 300 Hz, but M neurons could not follow even a low-frequency stimulus. Once the receptive field was determined, pressing with a blunt probe or a small cotton ball for pressure, forceps at 30–40°C for heat stimulation and chilled ones for cold, and a hypodermic needle for nociceptive stimulation were used to distinguish the other modalities.

Experimental Protocols

Depolarizing current applied through the recording electrode was increased by 0.1-nA steps until at least one action potential could be initiated. Threshold was defined as a 0.5 response probability. The R was calculated from the linear portion of the slope, which shows the current–voltage relationship between steady-state voltage deflection and E_m. Negativity of 20 mV or more was obtained when weak (<3-nA) hyperpolarizing pulses of 40-msec duration were applied through the intracellular electrode. τ was measured from the exponential change of potential produced by a weak hyperpolarizing current pulse. C was calculated from the formula $C = \tau/R$ (Ito, 1957).

Statistical Analysis

Data are presented as means ± SD. Statistical significance between populations of data was tested using Student's t test. Differences in means were considered statistically significant at $P < 0.05$.

RESULTS

Twenty-four M neurons and 19 V + M neurons in the trigeminal ganglion were recorded intrasomally.. Mean resting membrane potentials (E_m) were −53.0 mV (±3.2 SD; $n = 24$) for M neurons and −57.0 mV (±3.6 SD; $n = 19$) for V + M neurons. The E_m's were significantly different ($P < 0.05$) but with overlapping. Their modality was identified as explained under methods.

The Receptive Fields

The receptive fields of M neurons were 1–2.5 mm in diameter ($n = 21$), and those of the V + M neurons generally a little larger (2–4 mm in diameter; $n = 16$). The boundary of both kinds of receptive field was clear, and the shape was usually round or ellipsoidal for both modalities. Both types of neurons had only one receptive field on the skin or mucosa. These two were clearly separate populations of sensory cells. There was no overlap. These neurons did not respond to other sensory stimuli such as heat or pain. These results were similar to others reported proviously (Terashima and Liang, 1994a).

Response to Sustained Depolarization

Depolarizing current was injected intrasomally and elicited a spike discharge for M and V + M neurons at the threshold (T) current intensity (Figs. 1A and 1B).

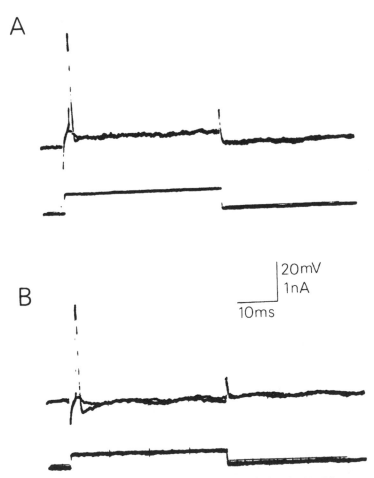

Fig. 1. Intrasomal recordings from an M neuron (A) and a V + M neuron (B) which responded to internal stimulation at threshold level. Two superimposed records demonstrate the all-or-nothing nature of the response. $E_m = -53$ mV in A and -57 mV in B. Calibration in B for both.

Table I. Passive Membrane Properties of M and V + M Neurons[a]

	M (n)	V + M (n)	
T (nA)	0.58 ± 0.21 (23)	0.3 ± 0.15 (19)	<0.001*
τ (msec)	0.86 ± 0.19 (19)	1.18 ± 0.37 (16)	<0.01*
R (MΩ)	13.6 ± 2.49 (19)	6.94 ± 2.04 (16)	<0.001*
C (pF)	64 ± 12 (19)	166 ± 33 (16)	<0.001*
Rebound latency (msec)	3.4 ± 0.7 (6)	4.1 ± 1.1 (9)	>0.05
Rebound T (NA)	0.70 ± 0.2 (6)	0.47 ± 0.11 (9)	<0.05*

[a] Values are means ± SD. Numerals in parentheses indicate the number of neurons. A two-tailed t test was used.
* A significant difference between means (i.e., $P < 0.05$).

We measured the AP threshold (T) values (Table I). T was higher in M neurons than in V + M neurons. However, like E_m, the difference between the average thresholds (0.6 ± 0.2 nA for M neurons, 0.3 ± 0.2 nA for V + M neurons) was too small to be useful in classifying cell types. Vibrotactile cells seemed to have a very low threshold (0.3 nA × 7 MΩ = 2.1 mV).

When we increased the depolarizing current intensity, 22 of 24 M neurons always responded with a single spike (Fig. 2A), while the remaining 2 responded like V + M neurons.

All 19 V + M neurons responded with multiple spikes to the suprathreshold stimulation current (Fig. 2B). The number of spikes increased with the depolarization (Fig. 3). The minimum interval (highest frequency) of V + M neurons increased with the current intensity, while the maximum intervals (lowest frequencies) remained relatively constant (Fig. 3).

Response to Sustained Hyperpolarization

When we applied a low hyperpolarizing current the time course of the change in the membrane potential at the onset of the pulse followed a simple exponential (Figs. 4A and B). We measured the time constant (τ) as in Figs. 4C and D. The data are presented in Table I. Although these data are significantly different, they were still too close (0.9 ± 0.2 msec for M and 1.2 ± 0.4 msec for V + M) to be used to differentiate cell type.

The input resistance (R) of the cells was determined, using a series of hyperpolarizing current injections (Fig. 4A for M and Fig. 4B for V + M). Then current stimulation graphs of the current–voltage relationship for M and V + M neurons were made (Figs. 5A and B). R was larger in M neurons than in V + M neurons (Table I). This membrane property shows the largest difference (13.6 ± 2.5 mV for M and 6.9 ± 2.0 mV for V + M cells) and thus the best potential for differentiating cell types.

We calculated C with a formula (see Methods) (Table I). C showed a large significant difference between the two sensory modalities: C is smaller in M neurons than in V + M neurons.

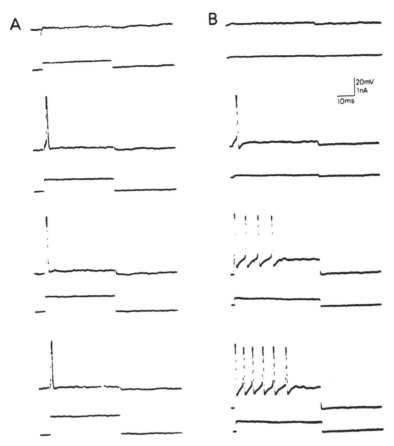

Fig. 2. Intrasomal recordings from an M neuron (A) and a V + M neuron (B) which responded to various prolonged depolarization. Upper trace, voltage response. E_m was −53 mV in A and −58 mV in B. Lower trace, injected current. Calibration in B for all.

Rebound potentials also appeared with a latency of less than 10 msec (Fig. 6). We measured the rebound latency and the rebound T (Table I). Rebound T was larger in M neurons than in V + M neurons.

DISCUSSION

Our newly obtained data were passive membrane properties including C, R, τ, and T, in addition to rebound T and rebound latency (Table I). We also compared some less studied electrophysiological properties, that is, the firing pattern of a soma in response to a long depolarizing current injection in M and V + M neurons (Figs. 2 and 3). The present observations on response properties to an adequate stimulus and receptive fields were consistent with those of a previous report (Terashima and Liang, 1994a).

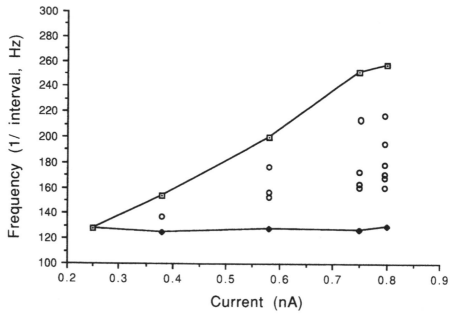

Fig. 3. The relationship between depolarizing current and discharge frequency (1/interspike interval) in a V + M neuron. The upper line connects frequencies of the first intervals (squares) for each current intensity, and the lower line connects frequencies of the last intervals (diamonds). Circles between the two lines represent intervals between the first and the last, which progressively increase in number for higher stimulus intensities and successively decrease in discharge frequency in the train for each stimulus intensity.

Current Stimulation and Adaptation Tendency

Although neither M nor V + M neurons have background discharges, their models of response were quite different (Terashima and Liang, 1994a). M neurons showed a train of spike discharges to a sustained mechanical stimulus applied on the receptive field, whereas V + M neurons showed a single discharge at the onset of stimulus. Repetitive firing during depolarization by injected current was related to modality, but the response was reversed: slowly adapting M neurons produced a single spike, and V + M neurons, which adapt rapidly to a sustained stimulus, produced multiple APs. When a spike train was elicited in an M neuron, slow depolarization might have been generated in the receptor, but the rapidly adapting tendency of this somal membrane property was not directly related to the receptor membrane properties. Martin and Wickelgren (1971) also concluded that there is dissimilarity between the dorsal neurons and receptors of the spinal cord in the lamprey. To find repetitive firing in A-β neurons seems a possible and convenient way to separate V + M from M neurons before searching for their receptive fields or during recording *in vitro*.

Passive Electrophysiological Properties

It has been reported that each type of dorsal cell has its own characteristic membrane properties [(Christenson *et al.*, 1988) lamprey]. The rapidly adapting

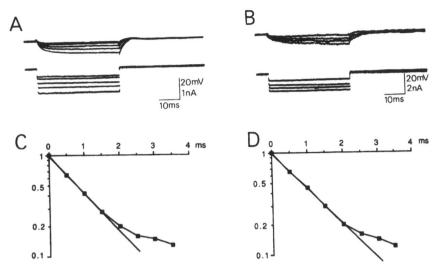

Fig. 4. Effects of prolonged hyperpolarization. The five and four superimposed records are those of an M neuron (A) and a V + M neuron (B), respectively. Upper trace, voltage response. E_m was −60 mV for A and −59 mV for B. Lower trace, injected current. The time course of the polarized potential in the rising phase of hyperpolarization is plotted for demonstration of the fitting procedure for an M neuron (C) and a V + M neuron (D). The membrane time constant was determined by plotting the time course of the change in membrane potential to a constant-current hyperpolarizing pulse. The straight line shows the initial time course, from which the ideal time constant was calculated (A, 1.1 msec; B, 1.2 msec).

cells in the lamprey (Martin and Wickelgren, 1971), which are reported to have a lower R, correspond to our V + M neurons. It seems reasonable that M neurons have lower τ, higher R, and lower C than V + M neurons (Table I), because the M soma is smaller than that of V + M neurons (Terashima and Liang, 1994a). Similar observations concerning the R and soma size relationships have been reported for rats (Harper and Lawson, 1985; Ritter and Mendell, 1992).

Most neurons with long AHP fire a single AP in response to prolonged deporalizing current (Gallego, 1983; Kostyuk et al., 1981). This suggests that prolonging the hyperpolarizing state is one of the mechanisms that limit a cell's firing. The works of Puil and Spigelman [(1988) guinea pig] and of Stansfeld et al. [(1986) rat] on sensory neurons, which show that blockade of a K^+ current facilitates firing, suggest that this difference in single versus repetitive spiking is due to a slowly inactivating K^+ current.

Although M and V + M neurons have similar amplitudes and durations of afterhyperpotential, CVs, and axon morphologies (Terashima and Liang, 1994a), they are quite different in spiking as well as in some passive electrophysiological properties of the somal membrane.

Action Potential Characteristics

Harper and Lawson (1985) reported that in the A-β group in the rat DRG there are two distinct neuron groups which are different in spike form: some have

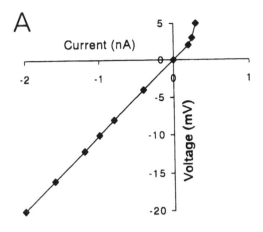

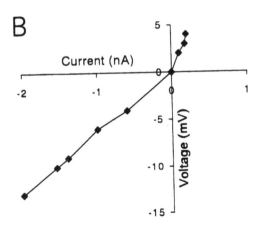

Fig. 5. The current–voltage relations obtained from an M neuron (A) and a V + M neuron (B).

a narrow spike without a hump (A-β_0), and others a wide spike with a hump (A-β_1). We could not find the latter in our crotaline A-β group encompassing both M and V + M neurons. Our V + M neurons were similar to Harper and Lawson's A-β_I group in R (6.9 and 7.4, respectively), although ours had no hump (present result; Terashima and Liang, 1994a). TRG is composed of cells of both neural crest and ectodermal placode origin, while DRG cells are of neural crest origin only (Weston, 1970). Differences in ganglion origin and species seem to be affecting our cell classification and comparison.

   In conclusion, the spiking difference between M neurons and V + M neurons, which respond to suprathreshold depolarizing current differently, seems to be specific to the sensory modality, although it does not directly relate to the receptor membrane properties. On the basis of passive electrophysiological properties, a higher R is a practical means to differentiate M from V + M neurons in the A-β group in crotaline snakes.

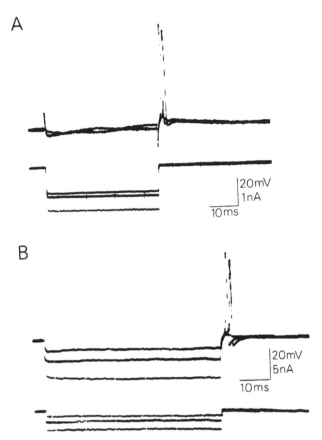

Fig. 6. Rebound responses of an M neuron (A) and a V + M neuron (B). Three superimposed traces for both. The first hyperpolarization current (highest traces) was under threshold level. Only the tail current appears. The second current trace was at threshold level (1 nA for A, 1.5 nA for B). The third (lowest) traces were elicited by suprathreshold current (1.4 nA for A, 3 nA for B). Compared with the rebound elicited by threshold current, the latency of rebound caused by the third current was shorter, and the amplitude was higher.

ACKNOWLEDGMENTS

We thank Dr. R. C. Goris for editing the English and Mr. Uehara for technical assistance in experiments and photography.

REFERENCES

Cameron, A. A., Leah, J. D., and Snow, P. J. (1986). The electrophysiological and morphological characteristics of feline dorsal root ganglion cells. *Brain Res.* **362**:1–6.

Christenson, J., Boman, A., Lagerbäck, P.-Å., and Grillner, S. (1988). The dorsal cell, one class of primary sensory neuron in the lamprey spinal cord. I. Touch, pressure but no nociception—A physiological study. *Brain Res.* **440**:1–8.

Fukuda, J., and Kameyama, M. (1980). A tissure-culture of nerve cells from adult mammalian ganglia and some electrophysiological properties of the nerve cells in vitro. Brain Res. **202**:249–255.

Gallego, R. (1983). The ionic basis of action potentials in petrosal ganglion cells of the cat. J. Physiol. (Lond.) **342**:591–602.

Harper, A. A., and Lawson, S. N. (1985). Electrical properties of rat dorsal root ganglion neurons with different peripheral nerve conduction velocities. J. Physiol. (Lond.) **359**:47–63.

Ito, M. (1957) The electrical activity of spinal ganglion cells investigated with intracellular microelectrodes. Jpn. J. Physiol. **7**:294–323.

Koerber, H. R., Druzinsky, R. E., and Mendell, L. M. (1989). Properties of somata of spinal dorsal root ganglion cells differ according to peripheral receptor innervated. J. Neurophysiol. **60**:1584–1596.

Kostyuk, P. G., Veselovsky, N. S., Fedulova, S. A., and Tysndrenko, A. Y. (1981). Ionic currents in the somatic membrane of rat dorsal root ganglion neurons. III. Potassium currents. Neuroscience **6**:2439–2444.

Liang, Y.-F., and Terashima, S. (1993). Physiological properties and morphological characteristics of cutaneous and mucosal mechanical nociceptive neurons with A-delta peripheral axons in the trigeminal ganglia of crotaline snakes. J. Comp. Neurol. **328**:88–102.

Liang, Y.-F., Terashima, S., and Zhu, A.-Q. (1995). Distinct morphological characteristics of touch, temperature, and mechanical nociceptive neurons in the crotaline trigeminal ganglia. J. Comp. Neurol. **360**:621–633.

Martin, A. R., and Wickelgren, W. O. (1971). Sensory cells in the spinal cord of the sea lamprey. J. Physiol. (Lond.) **212**:65–83.

Puil, E., and Spigelman, I. (1988). Electrophysiological responses of trigeminal root ganglion neurons in vitro. Neuroscience **24**:635–646.

Puil, E., Gimbarzevsky, B., and Miura, R. M. (1986). Quantification of membrane properties of trigeminal root ganglion neurons in guinea pigs. J. Neurophysiol. **55**:995–1016.

Puil, E., Gimbarzevsky, B., and Spigelman, I. (1988). Primary involvement of K^+ conductance in membrane resonance of root trigeminal ganglion neurons. J. Neurophysiol. **59**:77–89.

Ritter, A. M., and Mendell, L. M. (1992). Somal membrane properties of physiologically identified sensory neurons in the rat: Effect of nerve growth factor. J. Neurophysiol. **68**:2033–2041.

Rose, R. D., Koerber, H. R., Sedivec, M. J., and Mendell, L. M. (1986). Somal action potential duration differs in identified primary afferents. Neurosci. Lett. **63**:259–264.

Stansfeld, C. E., Marsh, S. J., Halliwell, J. V., and Brown, D. A. (1986). 4-Aminopyridine and dendrotoxin induce repetitive firing in rat visceral sensory neurones by blocking a slowly inactivating outward current. Neurosci. Lett. **64**:299–304.

Tanaka, H., Mishima, S., and Abe, Y. (1967). Studies on the behavior of Trimeresurus flavoviridis, a venomous snake on Amami Oshima Island in regard to speed of movement, nocturnal activity and sensitivity to infra-red radiation. Bull. Tokyo Med. Dent. Univ. **14**:79–104.

Terashima, S., and Liang, Y.-F. (1991). Temperature neurons in the crotaline trigeminal ganglia. J. Neurophysiol. **66**:623–634.

Terashima, S., and Liang, Y.-F. (1993). Modality difference in the physiological properties and morphological characteristics of the trigeminal sensory neurons. Jap. J. Physiol. **43** (Suppl. 1):S267–S274.

Terashima, S., and Liang, Y.-F. (1994a). Touch and vibrotactile neurons in a crotaline snake's trigeminal ganglia. Somatosens. Mot. Res. **11**:169–181.

Terashima, S., and Liang, Y.-F. (1994b). C mechanical nociceptive neurons in the crotaline snake's trigeminal ganglia. Neurosci. Lett. **179**:33–36.

Terashima, S., Zhu, A.-Q., and Chen, X.-Z. (1992). Modality specificity of membrane properties in ganglion neuron to internal stimulation. Neurosci. Res. **17** (Suppl. 17):S253.

Traub, R., and Mendell, L. M. (1988). The spinal projection of individual identified A-delta- and C-fibers. J. Neurophysiol. **59**:41–55.

Waddell, P. J., and Lawson, S. N. (1990). Electrophysiological properties of subpopulations of rat dorsal root ganglion neurons in vitro. Neuroscience **36**:811–822.

Weston, J. A. (1970). The migration and differentiation of neural crest cells. Adv. Morphog. **8**:41–114.

Yoshida, S., Matsuda, Y., and Samejima, A. (1978). Tetrodotoxin-resistant sodium and calcium components of action potentials in dorsal root ganglion cells of the adult mouse. J. Neurophysiol. **41**:1096–1106.

Research report

Distribution of NADPH-diaphorase in the central nervous system of an infrared-sensitive snake, *Trimeresurus flavoviridis*

Peng-Jia Jiang *, Shin-ichi Terashima

Department of Physiology, University of the Ryukyus School of Medicine, Nishihara-cho, Okinawa 903-01, Japan

Abstract

The distribution of NADPH-diaphorase (NADPH-d) activity was studied in the central nervous system of an infrared sensitive snake. An inhibitor of nitric oxide synthase, dichloroindophenol (DPIP), was used to distinguish the characteristics of NADPH-d activity. Intensely and weakly NADPH-d-stained neurons and fibers were found in discrete regions throughout the snake brain and cervical spinal cord, such as the olfactory bulb, subcommissural organ, stratum griseum periventriculare, locus coeruleus, dorsal root, dorsal horn, and area X. It was particularly noticed that the trigeminal descending nuclei and reticular formation of the medulla oblongata contained many positive neurons and fibers, but the lateral descending nucleus and nucleus reticularis caloris (infrared sensory nuclei) certainly did not. The positive neurons and fibers were also observed in supraspinal sensory ganglia. DPIP inhibited NADPH-d activity in all regions except for the olfactory/vomeronasal nerve and glomeruli. The results prove for the first time the presence of NADPH-d activity in the ophidian brain and suggest that nitric oxide may be involved in many neural functions, but not in infrared sensory processing.

Keywords: NADPH-diaphorase; DPIP; Enzyme histochemistry; Brain; Infrared; Snake

1. Introduction

Nitric oxide (NO) has received much attention as a newly classified neurotransmitter or neuromodulator [3,10,13,17,26]. This free radical can penetrate cell membrane freely by diffusion and modulate many neural functions. It is synthesized by NO synthase (NOS) from L-arginine. Mapping of this enzyme in the nervous systems may help us to understand the role of NO. Since reduced nicotinamide adenine dinucleotide-diaphorase (NADPH-d) has been proved to be a NOS [12,20], the identification and localization of NOS in various animals have been performed by NADPH-d histochemistry. NADPH-d activity in many regions has a distributions similar to that of NOS immunoreactivity. In the reptilia, NADPH-d-positive neurons in the turtle brain show much colocalization with NOS immunoreactive populations [7]. However, in some regions NADPH-d activity appears not to be coincident with NOS immunoreactivity, which is regarded as NOS-unrelated NADPH-d activity [19,38,47].

NOS widely exists not only in mammals (cats [31]; hamsters [11]; monkeys [18,35]; rats [1,42,45]) but also in

birds (chicks [5,6]; quail [33]), reptiles (lizards [27,34]; turtles [7]), fish (goldfish [9,44]; rainbow trout [36]; sunfish [28]; Atlantic salmon [19]), and even in insects (various invertebrates [16]; spider [30]). Except for many similarities of NOS distribution in these species, interspecies differences are present, even in very closely related species, like rats and hamsters; chickens and quail; turtles and lizards, the goldfish and Atlantic salmon. NOS is colocalized, although not fully, with various neurotransmitters and neuropeptides, such as acetylcholine, calcitonin gene-related peptide (CGRP), GABA, substance P, and tyrosine hydroxylase (TH) [1,11,27,43,46]. In order to examine NOS in the ophidian brain, we used a crotaline snake, which has not yet been studied. The identification of NOS in the snake central nervous system was done with NADPH-d histochemistry combined with reaction with a NOS inhibitor.

This snake possesses a pair of pit organs – infrared receptors. In its trigeminal ganglion, the infrared neurons, i.e. temperature neurons, are medium in size, smaller than tactile and vibrotactile neurons and larger than C nociceptive neurons [37]. In its brainstem, the lateral descending nucleus and nucleus reticularis caloris process only infrared information, whereas the common trigeminal de-

scending nuclei and reticular nuclei function for other
sensory modalities as in other animals [22,40]. This animal
provides us an ideal experimental model to investigate NO
role further in sensory modalities.

Preliminary results have been reported elsewhere [41].

2. Materials and methods

Ten crotaline snakes, *Trimeresurus flavoviridis*, of both
sexes weighing 120–380 g were used. After the animals
were anesthetized with halothane and paralyzed with pancuronium (0.6–0.8 mg/1 time, i.m.), additional ketamine
(40 mg/kg, i.m.) or pentobarbital sodium (40 mg/kg,
i.m.) was administered. Following artificial respiration for
30 min, the animals were perfused through the carotids of
both sides with cold (4°C) heparinized saline and fixative
that contained 2.5% paraformaldehyde and 1.5% glutaraldehyde in 0.1 M phosphate-buffered saline (PBS, pH
7.4). The brain, segments 1–10 of the cervical spinal cord,
and sensory ganglia (maxillo-mandibular, ophthalmic, and
craniocervical ganglia) were removed. The craniocervical
ganglia [21] were carefully separated into the proximal and
distal ganglia, which are homologous to the jugular-superior and nodose vagal ganglia of the mammals, respectively. Then the tissues were postfixed 1 h in the same
fixative and stored overnight in 10% to 30% sucrose in 0.1
M PBS for cryoprotection. Coronal sections (some were
sagittal or horizontal) were cut at 30–50 μm thickness on
a freezing microtome. Serial sections were collected in
PBS.

The free-floating sections were incubated in reaction
mixture which contained 1 mg/ml β-NADPH (Kojin Co.,
Japan), 0.2 mg/ml nitroblue tetrazolium (Sigma), and
0.3% Triton X-100 in 0.1 M PBS (pH 7.4) at 37°C for 1–3
h. The reaction was stopped by rinsing the sections with
cold 0.1 M PBS. Then the sections were mounted onto
chrome-gelatin coated glass slides, and dried overnight in
air. To avoid abolishment of the reaction product, the
slides were dehydrated in ethanol rapidly, and then mounted
with Entellan.

In some cases, alternate sections were used for different
reactions. (1) Histochemical control: the sections were
incubated in the reaction mixture without β-NADPH. (2)
NOS inhibitor: the sections were incubated in the reaction
mixture containing 0.01 mM or 0.1 mM dichloroindophenol (DPIP). As an artificial electron acceptor, DPIP inhibits NOS activity by taking the electrons which are
necessary for converting arginine to citrulline and NO
[23,24] and DPIP is able to compete with nitroblue tetrazolium for electrons, and thus can inhibit NADPH-d activity. We chose it because the L-arginine antagonists such as
L-monomethyl-L-arginine and N^ω-nitro-L-arginine do not
affect NADPH-d activity in fixed tissue [38].

The nomenclature used is that of Molenaar (*Python
reticulatus*, [32]) and Kusunoki et al. (*Agkistrodon
blomhoffi*, [25]).

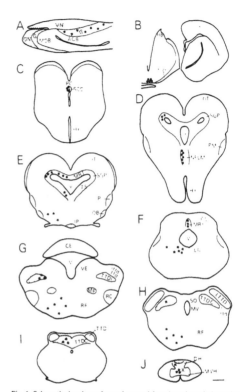

Fig. 1. Schematic drawings of rostral-to-caudal sections through the snake
brain and spinal cord showing the distribution of NADPH-d stained
elements. A is a horizontal section and B–J are coronal sections. Stars
indicate stained cells with their processes, large dots stained somata, and
small dots fibers and terminals. The number of the symbols does not
indicate a quantitative evaluation. Bar = 500 μm. Abbreviations in this
and following figures are: AOB, accessory olfactory bulb; CC, colliculus
caudalis; CE, cerebellum; CO, cortex; DH, dorsal horn; G, glomeruli;
HB, habenula; HY, hypothalamus; I, nucleus isthmi; IP, interpeduncular
nucleus; LC, locus coeruleus; LTTD, lateral descending nucleus of the
trigeminal nerve; MF, motor nucleus of the facial nerve; MOB, main
olfactory bulb; MRT, mesencephalic trigeminal nucleus; MV, motor
nucleus of the vagus nerve; MVH, ventral horn; NFLM, nucleus of
fasciculus longitudinalis medialis; OB, nucleus opticus basalis; ON,
olfactory nerve; OT, optic tectum; PC, posterior commissure; PL, nucleus
posterolateralis tegmentalis; PM, nucleus profundus mesencephali; RC,
nucleus reticularis caloris; RF, reticular formation; SAC, stratum album
centrale; SCO, subcommissural organ; SGP, stratum griseum periventriculare; SO, nucleus of the solitary fasciculus; TTDC, subnucleus caudalis;
TTDI, subnucleus interpolaris; TrO, optic tract; TS, laminar nucleus of
the torus semicircularis; V, ventricle; VE, vestibular nucleus; VN,
vomeronasal nerve; X, area X; lttd, lateral descending tract of the
trigeminal nerve; ttd, descending tract of the trigeminal nerve; r, motor
root of the trigeminal nerve.

3. Results

In the snake, NADPH-d-positive neurons and fibers were observed in discrete regions throughout the brain, spinal cord (Fig. 1), and sensory ganglia. The neurons were stained essentially in two patterns. In one pattern, only the somata were stained, for example, periglomerular cells and neurons in the nucleus of the fasciculus longitudinalis medialis and in the mesencephalic trigeminal nucleus (Fig. 2B; Fig. 3E, F). In the other pattern, intensely stained perikaryal cytoplasm with its processes and unstained nuclei was displayed, for example, small neurons in the stratum griseum periventriculare of the optic tectum, and neurons in the locus coeruleus, in the reticular formation, and in the subnucleus caudalis of the trigeminal descending nucleus (Fig. 3F; Fig. 4B, E; Fig. 5B). Bifurcations of the processes were found easily but not many varicosities. Almost no axons were identified. Stained glial cells were prominently seen in the brainstem and the spinal cord. No positive elements were found in the control reaction. All NADPH-d activity was inhibited by DPIP except in the olfactory/vomeronasal nerve and glomeruli. Notable regions of DPIP inhibition are explicitly described below.

3.1. Forebrain

In the olfactory bulbs, the olfactory/vomeronasal nerve and glomeruli were stained, with more intensity in the accessory bulb (Fig. 1A; Fig. 2A). Strongly stained small cells (4–8 μm) were found surrounding the glomeruli in the accessory olfactory bulb, but not in the main bulb. However, their processes and nuclei were unidentified except for densely stained somata (Fig. 2B). DPIP (0.01 mM) completely inhibited NADPH-d activity in such small cells, but not in the olfactory/vomeronasal nerve or glomeruli (Fig. 2C, D). No other positive neurons and fibers were found in either bulbs.

A group of moderately stained cell bodies in the supraoptic area was seen at the ventral edge of the rostral diencephalon and near the optic nerve (Fig. 1B; Fig. 3A). The paraventricular area was essentially not stained although sometimes 3–5 very lightly stained cells appeared.

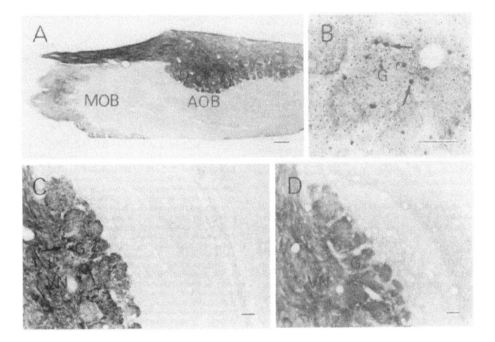

Fig. 2. NADPH-d-stained elements in the olfactory bulb. A: horizontal section showing positive olfactory/vomeronasal nerve fibers and glomeruli. B: periglomerular cells (arrows) in the accessory bulb. All the fine dots are reaction artifacts. C, D: alternate sections of NADPH-d histochemical reaction (c) and inhibitor reaction (0.01 mM DPIP, D). Note that the stained periglomerular cells (arrows) are inhibited by DPIP, but the vomeronasal nerve and glomeruli are not (D). Bars: A = 200 μm; B–D = 40 μm. MOB, main olfactory bulb; AOB, accessory olfactory bulb; G, glomeruli; VN, vomeronasal nerve.

We did not find NADPH-d-positive neurons or fibers in the wide region of the cortex (Fig. 1B).

3.2. Brainstem

In the rostral mesencephalon, the subcommissural organ was very heavily stained, particularly in the apical portion of the ependymal cells (Fig. 1C; Fig. 3B). Heavily stained perikaryal cytoplasm surrounded the unstained nuclei so that the organ looked like a honeycomb. There were some stained fibers toward the posterior commissure, but no axon bundles in the vicinity. Lower DPIP concentration (0.01 mM) reduced the staining intensity, and 0.1 mM inhibited all the staining (Fig. 3C). Other paraventricular organs and the wall of the ventricle did not display positive reaction. The nucleus profundus mesencephali contained intensely stained fibers (Fig. 1D). Some of them appeared to surround local capillaries (Fig. 3D). The nucleus of the fasciculus longitudinalis medialis was NADPH-d-positive, showing moderately stained cell bodies arranged in a line (Fig. 1D; Fig. 3E). The nucleus posterolateralis tegmentalis and opticus basalis were not stained, but a cluster of positive bipolar cells was found

medially to them, possibly comprising mesencephalic reticular neurons (Fig. 1E). The interpeduncular nucleus contained only moderately stained neuropil (Fig. 1E).

Within the optic tectum (Fig. 1E; Fig. 3F), neurons of mesencephalic trigeminal nucleus were stained moderately to weakly. A group of intensely stained small pyriform cells (7–12 μm) were observed conspicuously in the dorsal and ventral stratum griseum periventriculare but not between the two in the laminar nucleus of the torus semicircularis (Fig. 1E). Their dendritic processes were prominently stained, and extended laterally (Fig. 3F).

The medulla oblongata had abundant NADPH-d-positive structures. The locus coeruleus was prominently stained and a plexus of positive fibers was in the neighbourhood (Fig. 1F; Fig. 4A). Numerous stained neurons had much dense reaction product in their perikaryal cytoplasm which blurred the cellular nuclei in some neurons. The multipolar neurons had obvious processes which were close to local capillaries (Fig. 4B). NADPH-d activity here was inhibited completely by 0.01 mM DPIP (Fig. 4C). The nucleus isthmi, colliculus caudalis, and cerebellum did not display any positive reaction (Fig. 1F, G). Many moderately stained glial cells and intensely stained multipolar

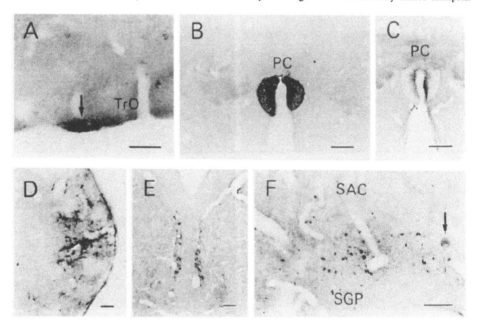

Fig. 3. NADPH-d-stained elements in the diencephalon and mesencephalon. A: sagittal section of the supraoptic area showing a cluster of cell bodies (arrow). B, C: alternate sections of the subcommissural organ stained with NADPH-d (B) which is inhibited by 0.1 mM DPIP (C). D: the nucleus profundus mesencephali. E: the nucleus of the fasciculus longitudinalis medialis. F: small pyriform neurons in the stratum griseum periventriculare of the optic tectum. An arrow indicates a moderately stained neuron of the mesencephalic trigeminal nucleus. Bars = 80 μm. TrO, optic tract; PC, posterior commissure; SAC, stratum album centrale; SGP, stratum griseum periventriculare.

neurons with bifurcated processes were scattered in a wide region of the reticular formation (Fig. 1G, H; Fig. 4E). But the nucleus reticularis caloris contained only stained glial cells without positive neurons or fibers (Fig. 1G; Fig. 4F). The neuropil of this nucleus appeared paler than surrounding area. All the motoneurons of the cranial nuclei were negative or very weakly stained (Fig. 1G, H; Fig. 4D).

Among the trigeminal sensory nuclei, the principal nu-

cleus and three subnuclei of the descending nucleus displayed moderate staining in the neuropil (Fig. 1G, H, I). Positive neurons were observed in the medial part of the subnucleus interpolaris and in the subnucleus caudalis (Fig. 1G, I). In the subnucleus interpolaris, a group of small neurons with prominent processes were particularly clearly stained (Fig. 5A), and the subnucleus caudalis, mainly in its caudal part, had many stained oval or round

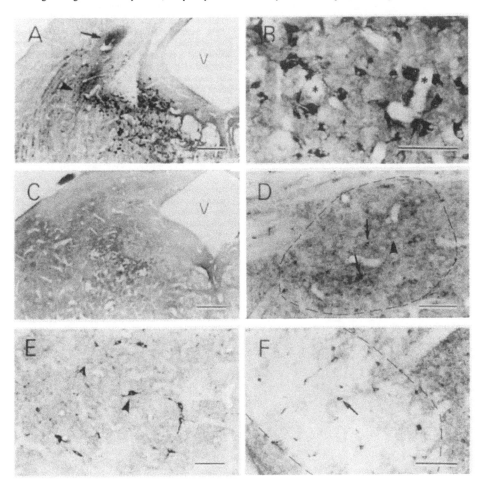

Fig. 4. NADPH-d-stained elements in the rhombencephalon. A: very strongly stained neurons in the locus coeruleus (arrowhead) and a plexus of fibers (arrow). B: high magnification of the locus coeruleus. Heavily stained neurons closely surround the capillaries (∗). C: NOS inhibitor reaction. DPIP (0.01 mM) completely inhibits the staining in the locus coeruleus (compare to Fig. 4A). D: the motor nucleus of trigeminal nerve (within the broken line). The motoneurons (arrowhead) are not clearly stained, but glial cells (arrows) are stained. E: the nucleus reticularis medius. Note intensely stained neurons with bifurcated processes (arrowhead). F: the nucleus reticularis caloris (within the broken line). Note stained glial cells (arrow) and pale neuropil. Bars: A = C = 200 μm; B, D, E, F = 80 μm. V, ventricle; r, motor root of the trigeminal nerve.

neurons with long fine processes that ran through the
nucleus to the tract (Fig. 5B, C). There were no stained
neurons or fibers in the lateral descending nucleus and its
tract (Fig. 1H, I; Fig. 5C). The stained fibers in the
descending tract never entered the lateral descending nucleus.

3.3. Cervical spinal cord

In segments 1–10 of the cervical spinal cord, longitudi-
nally and transversally cut NADPH-d-positive fibers were
dense in the dorsal root, Lissauer tract, and the dorsal
funiculus (Fig. 6A). Long positive fibers were seen in
horizontal sections (Fig. 6B). In the dorsal horn, there were
many intensely stained long thin fibers through areas
I–VII and some scattered neurons (Fig. 1J; Fig. 6A). In the
area near the central canal, area X, neurons with prominent
processes were heavily stained (Fig. 1J; Fig. 6C). The
motoneurons in the ventral horn were negative. In the
white matter, glial cells, some with their processes, were
moderately stained (Fig. 6D).

3.4. Sensory ganglia

We found stained cell bodies in all the supraspinal
sensory ganglia examined. In the trigeminal ganglia (Fig.
7A), positive cells in the ophthalmic and maxillo-mandibular ganglion displayed similar patterns and proportions.
The majority of the intensely stained cells were small.
Some medium-sized cells were moderately stained. Although we could not find their stem axons and bifurcations
satisfactorily, positive fibers were seen clearly in the
trigeminal roots and nerve trunks (Fig. 7B). They were
most abundant in the root, and more numerous in the
mandibular than in the maxillary nerve. The diameters
were about 1 μm or less. Of the craniocervical ganglia, all
the cells in the distal ganglion were heavily stained (Fig.
7D), while the proximal ganglion had only a small group
of stained cells located in the proximal portion (Fig. 7C).
Densely and weakly stained fibers were observed in the
vagus nerve trunk.

4. Discussion

The present results demonstrated NADPH-d-positive
neurons and fibers in the snake central nervous system,
suggesting the presence of NOS. Thus, the neurons and
fibers may synthesize and release NO, thereby modulating
cell functions in their targets. Although much of the
NADPH-d activity was inhibited by the NOS inhibitor
DPIP specific identification of NOS with immunocytochemistry would provide more secure results [8].

The overall distribution of NADPH-d-positive elements
in this snake appears to be similar to the other vertebrates
mentioned in the Introduction. However, interspecies differences were noted. As in the rodents and turtle olfactory
bulb [7,11,38], the olfactory/vomeronasal nerve,
glomeruli, and periglomerular cells in the accessory olfactory bulb of the snake are NADPH-d-positive and only the
staining of the periglomerular cells was inhibited by DPIP.
Since DPIP inhibits NOS activity as an electron acceptor,
whereas the olfactory/vomeronasal nerve and glomeruli

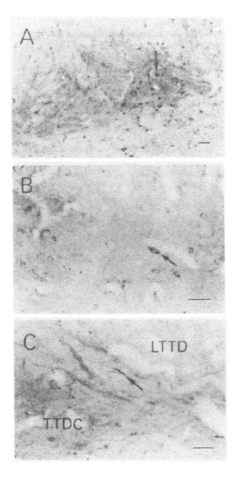

Fig. 5. NADPH-d-stained elements in the trigeminal sensory nuclei. A:
the subnucleus interpolaris. Note the densely stained neuropil and neurons
in the medial part (arrow). B: the subnucleus caudalis showing stained
neurons with long fine fibers. C: heavily stained fiber bundles (arrow) in
the tract of the trigeminal descending nucleus. Note that there are no
stained neurons or fibers in the lateral descending nucleus. Bars = 40
μm. LTTD, lateral descending nucleus of the trigeminal nerve; TTDC,
subnucleus caudalis.

had nonspecific staining, further immunocytochemical identification is required. In contrast, the periglomerular cells of the main bulb were NADPH-d-negative. Therefore, these results suggest that the role of NOS is different in the accessory and main olfactory bulbs of reptiles.

Nitroxergic neurons and fibers of the cortex and cerebellum appear to be prominently abundant in mammals [45] and are also found in lizards and turtles [7,34], and even in the teleost cerebellum [9]. However, they were absent in this snake species. It seems to represent a species which has eliminated expression of NOS from the cortex and cerebellum during its evolution. On the other hand, the strongly stained paraventricular organs are always observed in fish [28,36,44]. In the snake the subcommissural organ showed very dense staining of NADPH-d activity and the staining was completely inhibited by the higher concentration of DPIP (0.1 mM). However, in turtles and chickens this organ does not display NOS immunoreactivity [6,7]. This may suggest that NOS has a function in the snake subcommissural organ similar to that in fishes [28,36,44]. The locus coeruleus is one of the most interesting regions. Nitroxergic neurons in the area of the locus coeruleus have been identified in mammals except for the rat, and in birds, lizards, and turtles. These nitroxergic neurons have been strongly suggested to be coexpressed in cholinergic neurons [7,27,46]. This seems to hold true also in snakes. We found a dense accumulation of NADPH-d-positive neurons in the locus coeruleus, and Kusunoki et al. [25] have histochemically shown a strong acetylcholinesterase activity in the same nucleus of the snake.

As stated in Section 3, in the trigeminal nuclei of this snake the lateral descending nucleus had no NADPH-d-positive elements, which is completely contrary to the intensely stained common descending nuclei, particularly the subnucleus caudalis which contained many positive neurons and fibers. Similarly to the lateral descending nucleus, the nucleus reticularis caloris also did not contain positive neurons or fibers. Thus, it is reasonable to assume that NO is not related to information processing for infrared sense, but is involved in other somatosensory modalities. Unlike NOS, substance P-like immunoreactive (SP-LI) fibers have proved to be involved not only in the common trigeminal system but also in the infrared sensory system [21,39]. Therefore, the subpopulation of SP-LI fibers that is related to infrared sensation probably does not contain NOS.

The distribution of NADPH-d-positive neurons and fibers in the snake cervical spinal cord is quite similar to that in the mammals [2,4,14,15,43,47] and also in fish [9]. We observed that NADPH-d-positive neurons and fibers

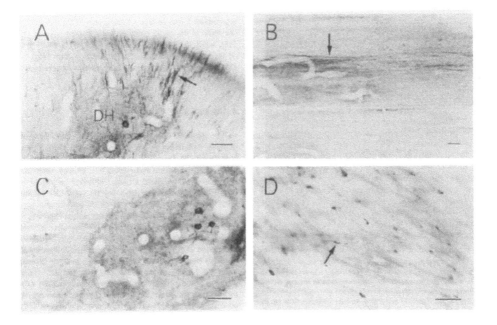

Fig. 6. NADPH-d-stained elements in the spinal cord. A: the dorsal horn, dorsal root, and dorsal funiculus. Note the many stained fibers (arrow). B: horizontal section showing long stained fibers (arrow). C: area X. D: stained glial cells (arrow) in the white matter. Bars = 40 μm. DH, dorsal horn.

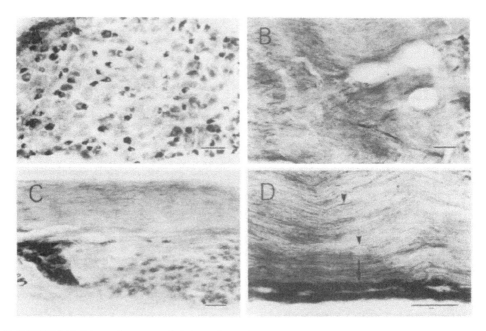

Fig. 7. NADPH-d-stained elements in sensory ganglia. A: the maxillo-mandibular ganglion. Note the many small positive cells. B: central root of the trigeminal ganglia. Note the many stained fibers running in parallel and cross ways. C, D: the proximal and distal craniocervical ganglia, respectively. Note that only part of the cells in the proximal ganglion are stained (C), whereas all the cells in the distal ganglion are stained (arrow in D). There are positive fibers (arrowheads in D) in the nerve trunk. Bars = 40 μm.

were concentrated in the dorsal root, dorsal horn, and area X, but absent in the ventral horn of the snake. NO also seems to modulate sensory processing in selected spinal regions of the snake.

This is the first evidence of NADPH-d-positive neurons and fibers in the reptilian sensory ganglia. Of the trigeminal nerve, the ophthalmic and maxillo-mandibular ganglia exhibited similar patterns. Intensely stained cells were mainly small ones. On the basis of cell size, this NADPH-d-positive subpopulation may be the same as the population of CGRP-positive cells that has been identified and quantitatively analyzed in the snake trigeminal ganglia of the same species in detail [37]. The coexistence of NOS containing cells with CGRP cells has been found in the rat ganglia [1]. As the case of CGRP, NO may also be a transmitter of small nociceptive neurons [29] but not of the infrared neurons. Therefore, this finding in the trigeminal ganglia is in accord with that in the trigeminal sensory nuclei. In the craniocervical ganglia, neurons in the distal ganglion were entirely NADPH-d-positive whereas those in the proximal one were only partially stained. This is similar to the case in the rat [1]. Nitroxergic neurons in the

snake distal ganglion may have similar functional role to the nodose vagal ganglion in the rat.

Glial cells are usually negative in other animals [33,45] or do not show NOS immunoreactivity even if they are NADPH-d-positive [19]. In the snake, we found moderately stained glial cells in the brainstem and spinal cord which were inhibited by DPIP. It is difficult to determine whether or not the glial cells in the snake contain more NADPH-d activity than in other species.

In conclusion, the present study is the first to suggest the presence of NOS throughout the snake brain, spinal cord, and ganglia. The distribution of NADPH-d activity appears to be similar to, but somewhat different from, that in other animals. This may reflect an evolutionary aspect of the nitroxergic system in the neural system. The facts that NADPH-d activity was positive neither diffusely in the central nervous system nor restrictedly within a specific functional pathway and that NOS seems to be partially colocalized with other neuroactive substances indicates a complexity and diversity of the NO role in the ophidian nervous system. NO may be involved in many functions in the ophidian nervous system, including somatosensory processing, but it seems not to be related to infrared sensory processing.

References

[1] Aimi, Y., Fujimura, M., Vincent, S.R. and Kimura, H., Localization of NADPH-diaphorase-containing neurons in sensory ganglia of the rat, *J. Comp. Neurol.*, 306 (1991) 382–392.

[2] Anderson, C.R., NADPH diaphorase-positive neurons in the rat spinal cord include a subpopulation of autonomic preganglionic neurons, *Neurosci. Lett.*, 139 (1992) 280–284.

[3] Bredt, D.S., Hwang, P.M. and Snyder, S.H., Localization of nitric oxide synthase indicating a neural role for nitric oxide, *Nature*, 347 (1990) 768–770.

[4] Brüning, G., Localization of NADPH diaphorase, a histochemical marker for nitric oxide synthase, in the mouse spinal cord, *Acta Histochem.*, 93 (1992) 397–401.

[5] Brüning, G., Localization of NADPH-diaphorase in the brain of the chicken, *J. Comp. Neurol.*, 334 (1993) 192–208.

[6] Brüning, G., Funk, U. and Mayer, B., Immunocytochemical localization of nitric oxide synthase in the brain of the chicken, *NeuroReport*, 5 (1994) 2425–2428.

[7] Brüning, G., Wiese, S. and Mayer, B., Nitric oxide synthase in the brain of the turtle *Pseudemys scripta elegans*, *J. Comp. Neurol.*, 348 (1994) 183–206.

[8] Brüning, G., NADPH diaphorase is not inhibited by ethylenediamineteraacetic acid and is not specific for nitric oxide synthase in the choroid plexus of rat and mouse, *Neurosci. Lett.*, 185 (1995) 16–19.

[9] Brüning, G., Katzbach, R. and Mayer, B., Histochemical and immunocytochemical localization of nitric oxide synthase in the central nervous system of the goldfish, *Carassius auratus*, *J. Comp. Neurol.*, 348 (1995) 183–206.

[10] Bult, H., Boeckxstaens, G.E., Pelckmans, P.A., Jordaens, F.H., Van Maercke, Y.M. and Herman, A.G., Nitric oxide as inhibitory non-adrenergic non-cholinergic neurotransmitter, *Nature*, 345 (1990) 346–347.

[11] Davis, B.J., NADPH-diaphorase activity in the olfactory system of the hamster and rat, *J. Comp. Neurol.*, 314 (1991) 493–511.

[12] Dawson, T.M., Bredt, D.S., Fotuhi, M., Hwang, P.M. and Snyder, S.H., Nitric oxide synthase and neuronal NADPH diaphorase are identical in brain and peripheral tissues, *Proc. Natl. Acad. Sci. USA*, 88 (1991) 7797–7801.

[13] Dawson, T.M., Dawson, V.L. and Snyder, S.H., A novel neuronal messenger molecule in brain: The free radical, nitric oxide, *Ann. Neurol.*, 32 (1992) 297–311.

[14] Dun, N.J., Dun, S.L., Förstermann, U. and Tseng, L.F., Nitric oxide synthase immunoreactivity in rat spinal cord, *Neurosci. Lett.*, 147 (1992) 217–220.

[15] Dun, N.J., Dun, S.L., Wu, S.Y., Förstermann, U., Schmidt, H.H.H.W. and Tseng, L.F., Nitric oxide synthase immunoreactivity in the rat, mouse, cat and squirrel monkey spinal cord, *Neuroscience*, 54 (1993) 845–857.

[16] Elofsson, R., Carlberg, M., Moroz, L., Nezlin, L. and Sakharov, D., Is nitric oxide (NO) produced by invertebrate neurons?, *NeuroReport*, 4 (1993) 279–282.

[17] Gally, J.A., Read Montague, P., Reeke Jr., G.N. and Edelman, M., The NO hypothesis: possible effects of a short-lived, rapidly diffusible signal in the development and function of the nervous system, *Proc. Natl. Acad. Sci. USA*, 87 (1990) 3547–3551.

[18] Hashikawa, T., Leggio, M.G., Hattori, R. and Yui, Y., Nitric oxide synthase immunoreactivity colocalized with NADPH-diaphorase histochemistry in monkey cerebral cortex, *Brain Res.*, 641 (1994) 341–349.

[19] Holmqvist, B.I., Östholm, T., Alm, P. and Ekström, P., Nitric oxide synthase in brain of a teleost, *Neurosci. Lett.*, 171 (1994) 205–208.

[20] Hope, B.T., Michael, G.J., Knigge, K.M. and Vincent, S.R., Neuronal NADPH diaphorase is a nitric oxide synthase, *Proc. Natl. Acad. Sci. USA*, 88 (1991) 2811–2814.

[21] Kadota, T., Kishida, R., Goris, R.C. and Kusunoki, T., Substance P-like immunoreactivity in the trigeminal sensory nuclei of an infrared-sensitive snake, *Agkistrodon blomhoffi*, *Cell Tissue Res.*, 253 (1988) 311–317.

[22] Kishida, R., Amemiya, F., Kusunoki, T. and Terashima, S., A new tectal afferent nucleus of the infrared sensory system in the medulla oblongata of crotaline snakes, *Brain Res.*, 195 (1980) 271–279.

[23] Klatt, P., Heinzel, B., John, M., Kastner, M., Böhme, E. and Mayer, B., Ca^{2+}/calmodulin-dependent cytochrome C redutase activity of brain nitric oxide synthase, *J. Biol. Chem.*, 267 (1992) 11374–11378.

[24] Klatt, P., Schmidt, K., Uray, G. and Mayer, B., Multiple catalytic functions of brain nitric oxide synthase. Biochemical characterization, cofactor-requirement, and the of N^{ω}-hydroxy-L-arginine as an intermediate, *J. Biol. Chem.*, 208 (1993) 14781–14787.

[25] Kusunoki, T., Kishida, R., Kadota, T. and Goris, R.C., Chemoarchitectonics of the brainstem in infrared sensitive and nonsensitive snakes, *J. Hirnforsch.*, 28 (1987) 27–43.

[26] Lowenstein, C.J. and Snyder, S.H., Nitric oxide, a novel biologic messenger, *Cell*, 70 (1992) 705–707.

[27] Luebke, J.I., Weider, J.M., McCarley, R.W. and Greene, R.W., Distribution of NADPH-diaphorase positive somata in the brainstem of the monitor lizard *Varanus exanthematicus*, *Neurosci. Lett.*, 148 (1992) 129–132.

[28] Ma, P.M., Tanycytes in the sunfish brain: NADPH-diaphorase histochemistry and regional distribution, *J. Comp. Neurol.*, 336 (1993) 77–95.

[29] Meller, S.T. and Gebhart, G.F., Nitric oxide (NO) and nociceptive processing in the spinal cord, *Pain*, 52 (1993) 127–136.

[30] Meyer, W., NADPH diaphorase (nitric oxide synthase) in the central nervous system of spiders (Arachnida: Araneida), *Neurosci. Lett.*, 165 (1994) 105–108.

[31] Mizukawa, K., Vincent, S.R., McGeer, P.L. and McGeer, E.G., Distribution of reduced-nicotinamide-adenine-dinucleotide-phosphate-diaphorase-positive cells and fibers in the cat central nervous system, *J. Comp. Neurol.*, 279 (1989) 281–311.

[32] Molenaar, G.J., The rhombencephalon of python reticulatus, a snake possessing infrared receptors, *Netherlands J. Zool.*, 27 (1977) 133–180.

[33] Panzica, G.C., Arévalo, R., Sánchez, F., Alonso, J.R., Aste, N., Viglitti-Panzica, C., Aijón, J. and Vázquez, R., Topographical distribution of reduced nicotinamide adenine dinucleotide phosphate-diaphorase in the brain of the Japanese quail, *J. Comp. Neurol.*, 342 (1994) 97–114.

[34] Regidor, J. and Poch, L., Histochemical analysis of the lizard cortex: an acetylcholinesterase, cytochrome oxidase and NADPH-diaphorase study. In W.K. Schwerdtfeger and W.J.A.J. Smeets (Eds.), *The Forebrain of Reptiles*, Karger, Basel, 1988, pp. 77–84.

[35] Sandell, J.H., NADPH diaphorase histochemistry in the macaque striate cortex, *J. Comp. Neurol.*, 251 (1986) 388–397.

[36] Schober, A., Malz, C.R. and Meyer, D.L., Enzymehistochemical demonstration of nitric oxide synthase in the diencephalon of the rainbow trout (*Oncorhynchus mickiss*), *Neurosci. Lett.*, 151 (1993) 67–70.

[37] Sekitani-Kumagai, M., Kadota, T., Goris, R.C., Kusunoki, T. and Terashima, S., Calcitonin gene-related peptide immunoreactivity in the trigeminal ganglion of *Trimeresurus flavoviridis*, *Neurosci. Res.*, 22 (1995) 287–295.

[38] Spessert, R., Wohlgemuth, C., Reuss, S. and Layers, E., NADPH-diaphorase activity of nitric oxide synthase in the olfactory bulb: Co-factor specificity and characterization regarding the interrelation to NO formation, *J. Histochem. Cytochem.*, 42 (1994) 569–577.

[39] Terashima, S., Substance P-like immunoreactive fibers in the trigeminal sensory nuclei of the pit viper, *Trimeresurus flavoviridis*, *Neuroscience*, 23 (1987) 685–691.
[40] Terashima, S. and Liang, Y.-F., Modality difference in the physiological properties and morphological characteristics of the trigeminal sensory neurons, *Jpn. J. Physiol.*, 43, Suppl. 1 (1993) S267–S274.
[41] Terashima, S. and Jiang, P.-J., Expression of NADPH-diaphorase in the crotaline snake nervous system, *Abstracts of SFN*, 21 (1995) p. 1619.
[42] Valtschanoff, J.G., Weinberg, R.J. and Rustioni, A., NADPH diaphorase in the spinal cord of rats, *J. Comp. Neurol.*, 321 (1992) 209–222.
[43] Valtschanoff, J.G., Weinberg, R.J., Kharazia, V.N., Schmidt, H.H.H.W., Nakane, M. and Rustioni, A., Neurons in rat cerebral cortex that synthesize nitric oxide: NADPH diaphorase histochemistry, NOS immunocytochemistry and colocalization with GABA, *Neurosci. Lett.*, 157 (1993) 157–161.
[44] Villani, L. and Guarnieri, T., Localization of NADPH-diaphorase in the goldfish, *Brain Res.*, 679 (1995) 261–266.
[45] Vincent, S.R. and Kimura, H., Histochemical mapping of nitric oxide synthase in the rat brain, *Neuroscience*, 46 (1992) 755–784.
[46] Vincent, S.R., NADPH-diaphorase histochemistry and neurotransmitter coexistence. In P. Panula, H. Päivärinta and S. Soinly (Eds.), *Neurohistochemistry: Modern Methods and Applications*, Alan R. Liss, New York, 1986, pp. 375–396.
[47] Vizzard, M.A., Erdman, S.L., Roppolo, J.R., Förstermann, U. and de Groat, W.C., Differential localization of neuronal nitric oxide synthase immunoreactivity and NADPH-diaphorase activity in the cat spinal cord, *Cell Tissue Res.*, 278 (1994) 299–309.

The Ultrastructure of Infrared Receptors in a Boid Snake, *Python regius*: Evidence for Periodic Regeneration of the Terminals

Fumiaki AMEMIYA[1], Richard C. GORIS[1], Yoshitoshi ATOBE[1], Norihisa ISHII[2], and Toyokazu KUSUNOKI [1]

1) Department of Anatomy, Yokohama City University School of Medicine, Fukuura 3-9, Kanazawa-ku, Yokohama 236.
2) Department of Dermatology, Yokohama City University School of Medicine, Fukuura 3-9, Kanazawa-ku, Yokohama 236.

SUMMARY

We did a light and electron microscope study of the infrared receptors in the labial pits of a python, *Python regius*. The receptors are elongated, tapering terminal nerve masses about 8 x 20 μ m located immediately below the conified surface of the pit fundus. The receptors are surrounded and supported by epithelial cells, and together with these epithelial cells are renewed and sloughed off periodically with each molting cycle. Each terminal nerve mass is packed with mitochondria in numbers not seen in any other sensory organ. Beneath the receptors is a dense network of fine capillary vessels which probably serve both to supply the tremendous energy requirements of the mitochondria-rich receptors and to stabilize and control the thermal sensitivity of the receptors. In active terminals (as opposed to degenerating ones), the mitochondria show both energized and discharged states, as well as ladder-like   junctions between individual mitochondria, which suggest that the mitochondria are intimately involved in producing the nerve impulse.

Key Words: Infrared receptors, Boid snakes, Mitochondria, Regeneration, Electron Micrographs.

Introduction

Boid and crotaline snakes possess infrared receptors which supplement vision in the detection and capture of prey (see review by Molenaar[12]; also Kobayashi et al.[8]). Many studies have been done on the physiology, innervation, and central projections of these receptors, but relatively little work has been done on their ultrastructure. In the work that has been done, the receptors have been shown to consist of terminal nerve masses (TNMs) that are densely packed with mitochondria[14,15]. These terminal nerve masses can be demonstrated by staining for the succinate dehydrogenase (SDH) contained in the mitochondria[4].

In the boids, the infrared receptors may be present under the surface of the labial scales, or they may be arrayed at the bottom of pits along the upper or lower lips, or along both upper and lower lips. These pits may be in the labial scales or between the labial scales (for a review, see Maderson[10]).

There has been one fairly thorough ultrastructural study of a pitless boid, *Boa constrictor*[15], but none of the seemingly more advanced boids with labial pits except for a short appendix by Meszler[11]. To remedy this lacuna, we undertook a light and electron microscope study of the labial pits of a small python, *Python regius*. As a result of this investigation, we discovered a hitherto unreported and possibly never before observed phenomenon: the infrared receptors are completely renewed with each molting cycle.

Materials and methods

We used two adult and three juvenile ball pythons, *Python regius*. The adults were about 100 cm in length and 1 kg in weight, and the juveniles about 50 cm in length and 55 g in weight. We used the two adults and two of the juveniles for electron microscopy. By coincidence, both adults were in the middle of the molting cycle, and of the two juveniles, one had just molted, and the other was in the final stages of the cycle. Each animal was anesthetized with halothane and perfused through the right aortic arch with heparinized (5 U/ml) 0.9% saline and 0.02% of an additional anesthetic, tricaine methanesulfonate, followed by perfusion with 2% paraformaldehyde and 2.5% glutaraldehyde in 0.1 M phosphate buffer at pH 7.4. The labial pits were then removed, and postfixed in 2% osmium tetroxide. After dehydration with ethanol, the labial pits were embedded in a mixture of Epon and Araldite or in Luft's Epon mixture. One micrometer semithin sections stained with toluidine blue were used for histologic examination and for selecting typical areas for ultrathin sections. The ultrathin sections were then stained for electron microscopy with uranyl acetate and lead citrate.

For the demonstration of SDH the remaining juvenile was anesthetized with halothane and then perfused through the right aortic arch with 200 ml of 0.9% saline solution containing 5 IU/ml heparin and 0.02% of tricaine methanesulfonate. Then the labial pits were dissected out and frozen, cross-sectioned serially with a cryostat, and mounted on gelatinized glass slides. The sections were dried at room temperature with a fan, and stained by the method of Nachlas[13] with various incubation times of from 3 to 6 minutes at 37℃.

Results

SDH staining

Six-minute SDH staining revealed a very darkly stained layer at the bottom of the pit organ. This was the layer containing the receptor terminals, and the heavy staining was indicative of the abundance of mitochondria contained therein (Fig. 1A). Also stained, but more lightly, were the unmyelinated nerve fibers (as identified in semithin and ultrathin sections and seen to contain large numbers of mitochondria, but fewer than the terminals) that fed into the receptor layer (Fig. 1A). There was no apparent SDH staining in the side walls of the pit.

In sections with a shorter incubation time (Fig. 1B) the terminals could be seen to form triangular masses with the base facing outward (i.e., toward the source of infrared rays). The triangular nerve masses were separated by numerous capillary blood vessels which appeared as unstained portions in the sections. (Red blood corpuscles also contain SDH, but the capillaries were empty, but not collapsed, because of the perfusion method.)

Semithin sections

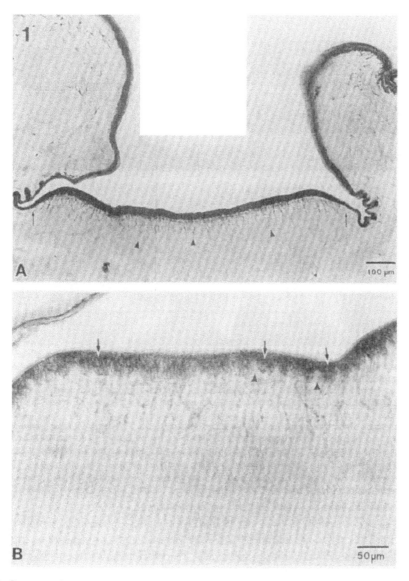

Fig. 1. Cross section through a python pit, stained for succinate dehydrogenase (SDH) in the mitochondria of the infrared receptors and their nerve fibers.

A: Arrows point to the right and left boundaries of the SDH staining in the fundus of the pit, which is the layer containing the terminal nerve masses (TNMs). This layer is deep purple in the actual sections. The staining on the vertical sides of the pit is due to melanin, and is a light brown in color. The unmyelinated nerve fibers leading to the receptors (arrowheads) are also stained because of the large number of mitochondria they contain.

B: A detail of the receptor layer shown in A, also stained for SDH, but for a shorter time, to reveal the inverted triangular shape of the TNMs (arrows). The nearly unstained areas (arrowheads) are the capillaries that run in the spaces between the apexes of the triangles.

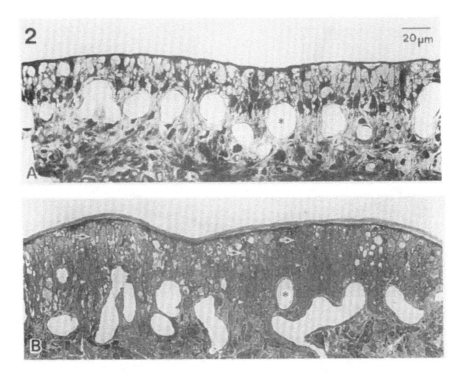

Fig. 2. Semithin sections through the receptor layer in 2 specimens at different stages of the molting cycle. Toluidine blue staining. The bed of dense, regularly spaced capillaries (asterisks) is characteristically apparent.

A: A specimen fixed immediately after shedding its skin. The receptor layer (measured from the cornified surface to the basement membrane) contains only well-formed, active TNMs, and is about 25 μ m thick.

B: A specimen fixed during the last stages of the molting cycle. The receptor layer is about 50 µm thick, and contains clearly degenerating TNMs (arrows, as judged from ultrathin sections in electron microscopy) in its outer half and apparently new, active TNMs in its inner half (arrowheads).

The semithin sections gave a good overview of the infrared receptor system (Fig. 2). Three main parts could be distinguished, which were, from the outside in, as follows.

1) Elongate masses of terminal receptors (the terminal nerve masses) clustered just beneath the outer cornified layer. The TNMs were interspersed with many epithelial cells which separated the TNMs from each other and appeared to have a supporting function. The TNMs themselves were elongate structures with the long axis orientated perpendicularly to the cornified surface and apparently in direct contact with it.

2) A dense, closely spaced network of capillary blood vessels, 15-25 μ m in diameter. The capillaries were oriented mostly parallel to the cornified layer and thus at right angles to the TNMs. They were virtually in contact with the TNMs, separated from the receptors only by the thin basement membrane, or, in places, the membrane and thin portions of the epithelial cells. The capillaries were spaced at strikingly regular intervals of 15-20 µm.

3) A layer containing Schwann cells and unmyelinated nerve fibers. The nerve fibers ascend toward the TNMs through the spaces between the capillary vessels.

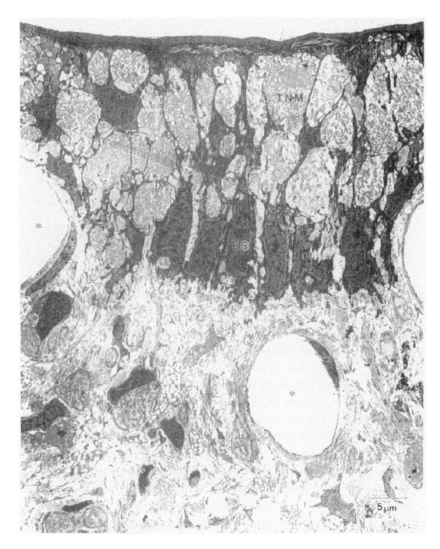

Fig. 3. An electron micrograph montage of ultrathin sections of the material shown in Fig. 2A. It presents a panoramic view of the receptor layer, showing the mitochondria-packed, triangular TNMs, the epithelial cells (EC) interspersed among and supporting the TNMs, and the extremely close proximity of the capillary bed (asterisks).

Electron micrographs

The electron micrographs confirmed all the above observations and provided further details (Fig. 3).

The cornified layer was about 2 μm thick. At high magnification, the surface of the layer contained tiny pits at fairly regular intervals (Fig. 4). The pits were about 0. 1 μm deep and 0. 15 μm wide, with nearly perpendicular sides.

The TNMs varied somewhat in size, but typically were 15-20 μm in long diameter and 5-8 μm in short diameter at the part in contact with the cornified layer, and tapering down toward the capillary bed (Fig. 3).

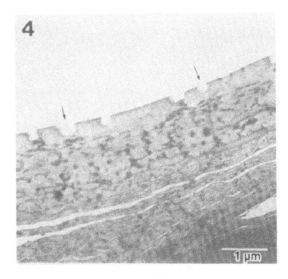

Fig. 4. A high magnification electron micrograph showing details of the cornified layer in the specimen of Fig. 2B. Arrows point to the tiny, regularly distributed pits described in the text.

All TNMs were densely packed with mitochondria, as suggested by the heavy SDH staining described above. The unmyelinated fibers leading to the terminals, which also stained lightly for SDH, also contained dense quantities of mitochondria, but less so than the TNMs.

There was a striking difference between material preserved just after sloughing (Fig. 3), and material preserved several weeks into the sloughing cycle (Fig. 5). In the latter material, supposedly active TNMs, as judged from the apparently normal configuration of their mitochondria, were located at some distance below the cornified layer (about 8 μ m in Fig. 5). Between them and the cornified layer could be seen degenerating terminals, with their mitochondria in various states of disintegration. The outermost (i.e., closest to the surface) terminals and mitochondria showed the greatest degree of disintegration (Fig. 5).

Some TNMs could be seen located within, and orientated at right angles to, other terminals (Fig. 6). This was not an isolated phenomenon, but could be seen in virtually all the material examined.

The normal-looking mitochondria in the active terminals presented various configurations. Some were twisted tubular ("energized") and others condensed ("discharged")[1] (see Fig. 7). Still others could be seen that were intermediate between these two extremes.

Junctions between individual mitochondria were also seen with a great degree of frequency. These were not simply juxtapositions, but the outer membranes of the mitochondria were connected by a series of structures that presented a ladder-like appearance in ultrathin sections (Fig. 8).

Just beneath the cornified layer we also observed a small number of clusters of clear vacuoles (Fig. 5) such as those interpreted by Jackson and Sharawy[7] as lipid droplets.

Discussion

Infrared reception is known to be present in all representative boid species, both those with and those without pits[1], but the relation of the infrared receptors with the sloughing cycle has never been addressed for

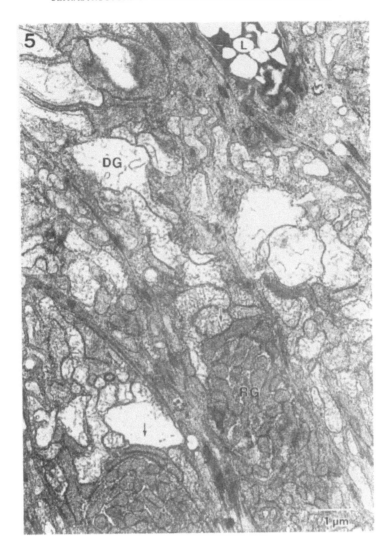

Fig. 5. The same as Fig. 4, but showing details of old, degenerating TNMs (DG) and newly regenerated, active TNMs (RG). The mitochondria in the old TNMs are clearly in a deteriorated state, swollen and crumbling interiorly, so that the demarcation between the old and the new TNMs (arrows) is immediately apparent. L: a typical cluster of lipid droplets.

any species, boid or crotaline, possessing infrared reception. In the boid specimens we used, some were perfused immediately after shedding (Fig. 3), and others were fixed at an intermediate stage in the sloughing cycle (Fig. 5). In those fixed at the intermediate stage, our material clearly shows receptors in at least 3 states, as judged from the condition of their mitochondria: active, functioning receptors in the deepest part of the epidermis, incipiently degenerating receptors above the active ones, and highly degenerated receptors in the process of becoming keratinized immediately beneath the outer keratinized layer.

As we have described in the results section, and as clearly indicated in the work of von Düring[15] as well, the infrared receptors are very evidently cutaneous receptors. Squamate reptiles shed the entire epidermis

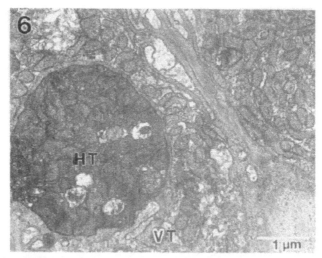

Fig. 6. The same as Fig. 4, but showing a TNM (HT) running horizontally inside and at right angles to a vertical TNM(VT).

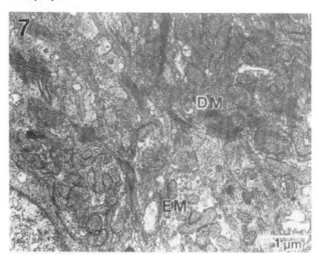

Fig. 7. The same as Fig. 4, but showing mitochondria in an energized (EM) and discharged (DM) state. The black spots are staining artifacts.

periodically, renewing each time all the cutaneous structures of the epidermis, including the receptor terminals included therein. This has been elegantly described and illustrated for touch receptors ("discoid receptors") in a gekkonid lizard, *Gekko gecko*[6]. Thus it is only reasonable to expect that the infrared cutaneous receptor terminals will also be renewed with each sloughing cycle, and this paper is the first to address this phenomenon.

The delineation between the active receptor terminals and the incipiently degenerating ones is very clear when one looks at the appearance of the mitochondria in these two types (Fig. 5). If this specimen had continued to live, it is to be expected that sloughing would have occurred at this point of delineation, once the

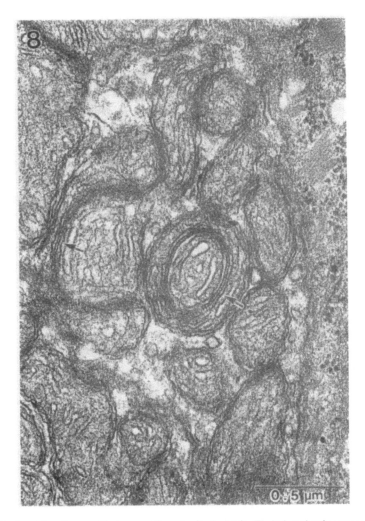

Fig.8 Ladder junctions (arrows) between adjacent mitochondria. Their function is as yet unknown.

keratinization of the degenerating terminals and other cutaneous structures was completed[9].

The tiny pits observed in the surface of the cornified layer (Fig. 4) were present in all the electron microscope preparations. They were apparent only in the surface above the receptor terminals, but not in other, non-receptive surfaces. We have observed the same surface architecture in scanning electron micrographs of the infrared receptor-containing labial scales in *Boa constrictor*, and of the crotaline pit membrane of *Agkistrodon blomhoffii* (unpublished). Therefore, this surface architecture appears to be an integral part of the infrared sensing system, since it is common to all types of snakes with infrared reception. We surmise that the array of pits serves as a sort of anti-reflection structure with regard to the infrared wavelengths, serving thus to sharpen the infrared image, just as coating does on optical lenses.

Changes in the configuration of the receptor terminal mitochondria at different phases of stimulation were first described by Meszler[11]. He described "energized" mitochondria in specimens fixed without any particular stimulation, and "discharged" mitochondria fixed immediately after an infrared stimulus. Thus it appears that mitochondria are intimately connected with the generation of the nerve stimulus.

In our material, we were able to see both "energized" and "discharged" mitochondria in the same material, even though it was not given any particular infrared stimulation before being fixed (Fig. 7).

However, peripheral infrared receptors in both crotalines and boids show constant spontaneous discharge, the average rate of which fluctuates with the ambient temperature, i.e., the temperature at which the receptors find themselves, assuming that the experimental animal, being a poikilotherm, has a body temperature equal to that of the ambient[2, 3, 5, 6].

Thus, if the mitochondria are indeed connected with the generation of nerve impulses, it appears only natural that energized and discharged mitochondria should be found simultaneously in material fixed at normally active body temperatures, because these receptors are constantly producing nerve impulses at such temperatures.

We also observed many cases of "junctions" between mitochondria. The ladder-like connections between the abutting mitochondrial membranes are very characteristic, and make it clear that here we are not dealing with a simple juxtaposition of organelles but with some form of mitochondrion-mitochondrion communication. Von Düring[15] also noticed this phenomenon and illustrated it, but made no comment on its significance. On the premise that the mitochondria are dynamically involved in producing the nerve impulse, these "ladder" junctions may be evidence of some form of dynamic communication between these organelles, such as the exchange of ions, etc. We found ladder junctions in almost all our material, both between mitochondria of similar states (energized-energized, discharged-discharged) and between mitochondria of dissimilar states (energized-discharged).

We found some receptor terminals oriented within and at right angles to other terminals (Fig. 6). This phenomenon has not been reported previously for any other snake with infrared reception, whether boid or crotaline. The precise configuration of these horizontally oriented terminals with regard to the vertical ones is difficult to judge from non-serial transmission electron micrographs, but such terminals appear with marked frequency in our material, regardless of the stage of the sloughing cycle (immediately after shedding, Fig. 3; several weeks after shedding, Fig. 5). The horizontal terminals are often more darkly stained than the surrounding vertical terminals, and frequently their mitochondria appear to be more tightly packed.

The function of the horizontal terminals in relation to the vertical ones is not immediately clear. It has been suggested that they serve to recognize the directionality of impinging infrared rays (H. D. Wolpert, pers. comm.). Whatever function they have, their very frequency of occurrence makes it clear that they are not an artifact of sectioning or staining.

Another characteristic of the boid infrared sensing system is the extensive system of capillaries underlying the layer containing the receptor terminals. Two principal functions can be attributed to this network. The first is obvious from the extreme density of the mitochondria packing the receptor terminals: these terminals have an extremely high energy budget, and therefore a high demand for oxygen. This is supplied by the capillary network, which is situated practically in contact with the terminals, permitting abundant and rapid transfer of energy. As a matter of fact, the network is so dense and the blood supply so abundant that the pits appear reddish to the naked eye in living specimens. Terashima and Goris noted in physiological experiments (unpublished) that central response to infrared stimulation ceased entirely whenever blood circulation in the pit membrane of crotalines became sluggish or stopped for some reason, such as a drop in blood pressure caused by too great a depth of anesthesia.

The second, and less obvious function of the capillary network can be considered to be temperature stabilization of the receptors. Physiological experiments in both boids and crotalines have shown that changes in the temperature of the receptors are responsible for initiating, sustaining, and coding their nerve impulses[2, 3, 6]. Excessive stimulation (i.e., an excessive rise in temperature) causes the receptors to shut down completely after an initial burst of firing[5]. Thus, for the receptors to function normally, some mechanism is necessary to carry away excess heat, not only during abnormal, "blinding" stimulation, but also during normal, physiological stimulation to return the receptor to its basal state as quickly as possible.

The capillary network seems eminently fitted for this function in virtue of its proximity to the receptor terminals. The capillary network will also serve to stabilize the receptors at a given level of background spontaneous firing by keeping them at the same temperature as the body temperature of the animal. It is also conceivable that changes in the rate of flow in the capillaries may serve to fine-tune the sensitivity of the receptors by regulating the time a stimulated receptor requires to return to its unstimulated baseline in readiness for the next, succeeding stimulus. In this way the blood flow in the capillaries would have a function somewhat analogous to that of the pupil of the eye in regulating sensitivity to visible light. In other words, the rate of blood flow serves to raise or lower the threshold of the potential generating mechanism.

Although somewhat more sophisticated, in being located in the fundus of a pit whose mouth can both serve as a pinhole lens and generate shadows whose edges move across the array of receptors in the fundus, the general ultrastructure of the receptors in these pits closely resembles that of the receptors in boids without pits. It is highly probable that in an evolutionary sense the infrared receptors in boids with pits are derived from those in boids without pits, whereas the infrared receptor organs in crotaline snakes appear to have evolved from some other lineage. This judgement is based on the fact that boid snakes are among the most primitive of snakes, retaining two lungs and vestiges of the pelvic girdle, whereas the crotaline snakes, with only a single lung and extremely sophisticated venom-injecting apparatus appear to be at the acme of ophidian evolution; yet infrared receptors are not present in any of the evolutionarily intermediate forms, such as the broadly diversified colubrids.

Acknowledgements

We wish to thank in particular Mrs. Chikako Usami and Mrs. Miki Kobayashi for their technical support.

References

1) Bullock, T. H. and Barrett, R. (1968): Radiant heat reception in snakes. *Comm. Behav.* Biol., (A) 1: 19-29.

2) Bullock, T. H. and Diecke, P. J. (1956): Properties of an infrared receptor. *J Physiol., Lond.*, **134**: 47-87.

3) de Cock Buning, Tj. , Terashima, S. and Goris, R. C. (1981): Crotaline pit organs analyzed as warm receptors. *Cell. Mol. Neurol.*, 1: 69-84.

4) Goris, R. C., Kadota, T. and Kishida, R. (1989): Innervation of snake pit organ membranes mapped by receptor terminal succinate dehydrogenase activity. *In* Current Herpetology in East Asia. (Matsui, M., Hidaka, T. and Goris R. C. ed.). 8-16. Herpetological Society of Japan, Kyoto.

5) Goris, R. C. and Nomoto, M. (1967): Infrared reception in oriental crotaline snakes. Comp. Biochem. *Physiol.*, **23**: 879-892.

6) Hensel, H. (1975): Static and dynamic activity of warm receptors in Boa constrictor. *Pfügers Arch.*, **353**: 191-199.

7) Jackson, M. K. and Sharawy, M. (1978): Lipid and cholesterol clefts in the lacunar cells of snake skin. *Anat. Rec.*, **190**: 41-46.

8) Kobayashi, S., Kishida, R., Goris, R. C., Yoshimoto, M. and Ito, H. (1992): Visual and infrared input to the same dendrite in the tectum opticum of the python, *Python regius*: electron-microscopic evidence. *Brain Res.*, **597**: 350-352.

9) Maderson, P.F.A. (1965): Histological changes in the epidermis of snakes during the sloughing cycle. *J Zool.*, **146**: 98-113.

10) Maderson, P. F. A. (1970): The distribution of specialized labial scales in the boidae. *In* Biology of the Reptilia, Vol. 2. (Gans, C. and Pearsons, T. S. ed.). 301-304. Academic Press, London and New York.

11) Meszler, R. M. (1970): Correlation of ultrastructure and function. *In* Biology of the Reptilia, Vol. 2. (Gans, C. and Pearsons, T. S. ed.). 305-314. Academic Press, London and New York.

12) Molenaar, G. J. (1992): Sensorimotor integration. In Biology of the Reptilia, Vol. 17, Neurology C. (Gans, C. and Ulinski, P. S. ed.). 367-453. The University of Chicago Press, Chicago.

13) Nachlas, M. M., Tsou, K. C. Souza, E., Cheng, C. S. and Seligman, M. (1957): Cytological demonstration of succinate dehydrogenase by the use of a new P-nitrophenyl substituted ditetrazol. *J Histochem. Cytochem.*, **5**:420-436.

14) Terashima, S., Goris, R. C. and Katsuki, Y. (1970): Structure of warm fiber terminals in the pit membrane of vipers. J. Ultrastruct. Res., **31**: 494-506.

15) von Düring, M. (1974): The radiant heat receptor and other tissue receptors in the scales of the upper jaw of Boa constrictor. *Z Anat. Entwickl.-Gesch.*, **145**: 299-319.

16) von Düring, M. and Miller, M. R. (1979): Sensory nerve endings of the skin and deeper structures. *In* Biology of the Reptiles, Vol. 9, Neurology A. (Gans, C., Northcutt, R. G. and Ulinski, P. ed.). 407-441. Academic Press, London.

Selective Labeling of [³H]2-Deoxy-D-Glucose in the Snake Trigeminal System: Basal and Infrared-Stimulated Conditions

Peng-Jia Jiang,[1] **and Shin-ichi Terashima**

Department of Physiology, University of the Ryukyus School of Medicine, Nishihara-cho, Okinawa 903-01, Japan

Abstract [³H]2-Deoxy-D-glucose (2-DG) and high-resolution autoradiography were employed to investigate labeling patterns of the trigeminal and infrared sensory system in a crotaline snake, the pit viper (*Trimeresurus flavoviridis*). Following intracardiac injection of 9.25 MBq [³H]2-DG, neurons in the nucleus of the lateral descending trigeminal tract (LTTD), nucleus reticularis caloris (RC), nucleus trigemini mesencephalicus, nucleus trigemini motorius, and trigeminal ganglia were labeled in various degrees after the pit organ had been removed (basal condition). This revealed that a higher rate of glucose utilization occurred in these nuclei than in the common sensory trigeminal nuclei, which lacked labeling entirely. When a pit was stimulated periodically with an infrared stimulus for 45 min, the difference in percentage of labeled cells was ipsilaterally increased by 12.84% in large cells of the LTTD and by 7.55% in the RC, as compared with the contralateral, basal-condition side. These slight changes indicate a small increase of glucose consumption during infrared reception. On the other hand, the small cells in the LTTD showed labeling that did not change with stimulation, suggesting that 2-DG uptake in inhibitory interneurons is relatively constant.

Key words high-resolution autoradiography, infrared pathway, nucleus descendens lateralis n. trigemini, nucleus reticularis caloris, neuron count, interneuron

Radioactive 2-deoxy-D-glucose (2-DG) labeling has been widely used for research on the functional activity and neuroanatomy of the central nervous system. In order to avoid diffusion of the soluble isotope (including 2-DG and 2-DG-6-phosphate) out of tissues, the traditional protocol uses frozen tissue sections to expose X-ray film or emulsion-coated slides after quick removal of the brain without perfusion fixation (Sharp et al., 1975; Sokoloff et al., 1977; Nudo and Masterton, 1986). However, findings in fine structures on X-ray film are masked, because the resolution is limited to about 100 μm. Cellular resolution with [³H]2-DG has been accomplished in several laboratories (Sejnowski et al., 1980; Durham et al., 1981; Pilgrim and Wagner, 1981; Sharp et al., 1993). The remaining insoluble labeling is identified after perfusion with fixative (Wolfe and Nicholls, 1967; Kai-Kai and Pentreath, 1981; Pentreath et al., 1982; Witkovsky and Yang, 1982; Nelson et al., 1984). Thus, 2-DG labeling patterns can be observed more conveniently and directly. Paraffin section emulsion autoradiography with [³H]2-DG has been recommended as a potential tool to provide a very precise comparison between particular nuclei

1. To whom all correspondence should be addressed.

and their functional correlates (Durham et al., 1981; Durham and Woolsey, 1985). Early studies demonstrated the patterns of cellular labeling corresponding to unfixed conventional methods, although loss of radioactivity was unavoidable during histological procedures (Witkovsky and Yang, 1982; Durham and Woolsey, 1985).

The pit organ, an infrared (IR) receptor in certain snakes, is innervated by the ophthalmic and supramaxillary trigeminal ganglia (Kishida et al., 1982). The IR sensory pathway consists of first-, second-, and third-order relays in crotaline snakes—that is, the ipsilateral nucleus descendens lateralis n. trigemini (LTTD), the nucleus reticularis caloris (RC), and the contralateral optic tectum (OT; Newman et al., 1980; Stanford et al., 1981). The large cells in the LTTD and the cells in the RC have been previously demonstrated to be IR relay cells through tracing with horseradish peroxidase (Newman et al., 1980). This pathway was studied with [¹⁴C]2-DG autoradiography by Auker et al. (1983; *Crotalus ruber, C. horridus, C. viridis helleri*) and Gruber et al. (1984; *Crotalus viridis*). In order to obtain more detailed information, we reexamined the 2-DG labeling pattern in this pathway at higher magnification with high-resolution [³H]2-DG autoradiography and paraffin embedding techniques. We were able to observe labeling characteristics that differed from those reported in previous studies. Preliminary results have been reported elsewhere (Terashima and Jiang, 1994).

MATERIALS AND METHODS

Animal Treatment

We used adult pit vipers (*Trimeresurus flavoviridis*) of both sexes, weighing 250–305 g and measuring 120–136 cm in length. These crotaline snakes possess a pair of pits, one on each side of the face. All animals were anesthetized with halothane and then ketamine (30–40 mg/kg, i.m.). After immobilization with pancuronium (2 mg/kg, i.m.), the snakes were maintained with artificial respiration (de Cock Buning et al., 1981). One or both pit organs of each animal were surgically removed 2.5 hr or 1 to 4 days prior to 2-DG injection. Both eyes were sutured shut either on the day the pit organ was removed or on the day of the experiment, and the snake was further blinded by Vaseline mixed with carbon powder. Before 2-DG injection, we waited 1–3 hr after halothane and pancuronium administration, and 10 min to 2 hr after ketamine administration. We used various intervals after operation or anesthesia to observe the effects of surgery and anesthesia on 2-DG uptake. They appeared not to have any effect on labeling patterns.

First, we examined the 2-DG labeling patterns under basal conditions in two animals with both pit organs removed. Each animal was intracardially injected with 9.25 MBq in 1 ml saline of $[1,2\text{-}^3H(N)]2\text{-}DG$ (sp. act. = 1.48 TBq/mmol; American Radiolabeled Chemicals Inc., St. Louis, MO, USA) in 1 min. After 45 min, transcardial perfusion was begun with cold Ringer's solution (NaCl 0.81%, KCl 0.022%, $CaCl_2$ 0.026%) with heparine and then 100 ml of periodate–lysine–paraformaldehyde fixative (pH 6.2), which fixes complex carbohydrates (McLean and Nakane, 1974). We added a glycolytic inhibitor, iodoacetic acid (0.02 M), to the fixative just before perfusion to prevent glycogenolysis (McCasland and Woolsey, 1988a).

Second, we examined the IR-stimulated labeling patterns in three animals, in which one pit organ was stimulated periodically with a standard IR stimulus (de Cock Buning et al., 1981) and the other pit organ was removed to provide basal conditions. A halogen lamp (100 V, 650 W) was adjusted to 15 V with a rheostat as a stimulus source and placed 30 cm from the opening of each snake's pit. This radiant source (energy flux = 19 mW/cm^2) was occluded by a thick cardboard disc with a coat of foil that had four pie-window cutouts. It was driven by a synchronous motor (8H25F-290; Japan Servo Co., Ltd., Tokyo, Japan). This made the radiant source sweep across the pit region for 1 sec every 3 sec. The head, except for the stimulated pit region, was covered with Styrofoam. IR stimulation was started just after isotope injection. Perfusion was begun 45 min after stimulation (2 hr in one case, but the results were

similar). 2-DG injection and perfusion procedures were the same as described above.

Autoradiographic Procedure

The brain and both trigeminal ganglia were immediately dissected after perfusion and postfixed for 1 hr in the same fixative solution. After dehydration in ethanol solution (50%, 70%, and 95%) and immersion in butanol (for 1 hr and then overnight), the tissues were embedded in paraplast. Serial coronal sections (8–10 μm) were cut and mounted on clean glass slides. The slides were dewaxed in fresh xylene, hydrated through graded alcohols to distilled water, dried overnight at 37°C, and dipped in Sakura NR-M2 autoradiographic emulsion (Konica Co., Tokyo, Japan; diameter of silver grains = 0.15 μm) diluted 1:2.5 with distilled water at 45°C. After 3–6 weeks of exposure in a refrigerator at 4°C, the emulsion-coated slides were developed in Konidol X (Konica Co.), fixed in Super Fujifix solution (Fuji Photo Film Co., Ltd., Tokyo, Japan), and counterstained with 0.1% thionin. Observations were made under a light microscope with darkfield and brightfield illumination (Olympus BH-2, Tokyo, Japan).

Data Collection

We counted the labeled cells to estimate uptake of $[^3H]2\text{-}DG$. All neuron counts were performed on every fourth section of 10-μm serial sections or every fifth section of 8-μm serial sections (total 40 μm in thickness). Because sometimes it was difficult to distinguish the cellular nucleus and nucleolus as a result of small cell size and obscuring by silver grains, neuronal bodies were counted in the LTTD, nucleus mesencephalicus n. trigemini (MRT), and nucleus motorius dorsalis n. trigemini (MTD), and cellular nuclei were counted in the RC, nucleus motorius ventralis n. trigemini (MT), and ganglia. No correction factor was employed because of the small diameter of neurons relative to section thickness (Konigsmark, 1970). In each brain, neurons were counted bilaterally, and data were represented in terms of percentages of labeled and unlabeled neurons from direct counting. In addition, the number and cross-sectional areas of labeled cells in the ophthalmic ganglia of three IR-stimulated animals were simultaneously measured with a Nikon three-dimensional analyzer system combined with a light microscope (Cosmozone 98; Nippon Kogaku K. K., Tokyo, Japan). We compared the two sides of the brain in each individual, and then compared the differences between basal and stimulated conditions.

To determine statistical differences, one-way analyses of variance were used for means of labeled cells and Kolmogorov–Smirnov tests for population distributions of labeled ophthalmic ganglion cells. Both types of tests were performed with a StatView 4.0 program on a Macintosh computer (Apple Computer, Inc., USA). The level of significance was set at $p < 0.05$.

RESULTS

[³H]2-DG Labeling under Basal Conditions

In seven half-brains from five animals, there were many labeled neurons in the MRT, the MT and MTD, the RC, and the LTTD after the pit organ was removed, but not in the common sensory trigeminal nuclei, including the nucleus sensorius principalis (SPT) and descendens n. trigemini (TTD). There was also heavy 2-DG labeling in the nucleus interpeduncularis. Large cells (Deiter's cells) in the nucleus vestibularis were also frequently labeled with the silver grains. The neuropil was lightly and homogeneously labeled by comparison with nearby perikarya.

THE TRIGEMINAL SYSTEM

Typical [³H]2-DG labeling was not found in the SPT and TTD, except for some glial cells and occasionally small neurons. Figures 1A and 1B show a part of the nucleus caudalis (TTDC) near the LTTD.

In two parts of the motor trigeminal nuclei, the MT and the MTD, the silver grains were densely distributed on the soma cytoplasm and proximal process, but were frequently absent from its nucleus. Some labeled linear elements could be typically observed under darkfield illumination (Fig. 2). The total percentage of labeled cells in these motor nuclei was high, nearly 90% or more, except in one case (Table 1).

The MRT was located in the periventricular gray layer of the OT. It consisted of spherical and dark Nissl-stained neurons $10-24$ μm in diameter ($n = 140$, Fig. 3). The majority of cells ($90.31\% \pm 0.51\%$ [SEM], $n = 7$) were densely labeled with 2-DG under basal conditions (Fig. 3, Table 1). The silver grains were over the cell bodies, mainly in the cytoplasm. The labeling could be reliably identified under darkfield observation.

THE IR SENSORY SYSTEM

The LTTD is the first-order nucleus in the IR sensory pathway. It is located in the dorsolateral medulla, surrounded by the nerve tracts of the LTTD and of the TTD. The range of cellular diameter in the LTTD is $7-27$ μm ($n = 828$). Following Meszler (1983) and Meszler et al. (1981), we also classified the cells as large ($15-27$ μm) and small ($7-14$ μm). The labeling was heavy in both the large and the small cells (Fig. 1). Some glial cells were also labeled. The mean percentage of labeled small cells ($69.68\% \pm 2.37\%$, $n = 7$) was a little higher than that of large cells ($48.31\% \pm 5.55\%$, $n = 7$) ($p < 0.05$, Table 1).

The RC, the second-order nucleus in the IR sensory pathway of the crotaline snake, is in the ventrolateral medulla. Medium-sized neurons ($11-30$ μm in diameter, $n = 302$) were scattered in the neuropil network. They were also labeled with 2-DG. The labeling of each cell in the RC seemed to be lighter than in the LTTD (Fig. 2), but the mean percentage of labeled cells ($46.87\% \pm 5.9\%$, $n = 7$) was similar to that of the large cells in the LTTD ($p > 0.05$, Table 1).

The third-order nucleus in the IR sensory pathway is considered to be in the middle lamina, stratum griseum centrale, of the OT (Auker et al., 1983; Gruberg et al., 1984; Northcutt, 1984). We did not find typical labeling in this lamina, apart from the radioactive background (Fig. 3).

[³H]2-DG Labeling after IR Stimulation

In three unilaterally IR-stimulated animals, the labeling patterns of the ipsilateral trigeminal nuclei in response to stimulus were essentially similar to those of the contralateral side under basal conditions (Fig. 4). No labeling was found in the contralateral middle lamina of the OT.

However, after IR stimulation, the total percentage of labeled large cells in the LTTD increased from 47.13% contralaterally (preparations 1, 2, and 3 in Table 1) to 59.97% ipsilaterally, and the total percentage of labeled cells in the RC increased from 40.46% contralaterally to 48.01% ipsilaterally. We quantitatively estimated the stimulation effect as the difference in change between the two sides of the brain (Fig. 5). In order to minimize uncontrollable variation, the animals with both pits removed as controls were compared to unilaterally IR-stimulated animals. The differences in percentage of labeled cells between the two sides of four nuclei were calculated in each case by subtracting the percentage of labeled cells on the contralateral side from that on the stimulated side in the IR-stimulated animals ($n = 3$). In the animals with both pits removed ($n = 2$), we also calculated the differnce between the two sides of the brain. The mean difference in percentage of labeled cells was prominently increased by 12.84% $\pm$ 2.15% in large cells of the LTTD, and 7.55% $\pm$ 0.72% in the RC on the IR-stimulated side as compared with the contralateral side. These increases were significant ($p < 0.05$) as compared to control levels (2.42% $\pm$ 0.94% and 2.49% $\pm$ 0.41%, respectively). However, the changes in labeled small cells in the LTTD, and labeled cells in the MRT and in the trigeminal motor nuclei, were not significantly different between the IR-stimulated experiments and the controls ($p > 0.05$).

[³H]2-DG Labeling in the Trigeminal Ganglia

About 30% of the neurons in the ophthalmic and the maxillomandibular trigeminal ganglia were labeled with 2-

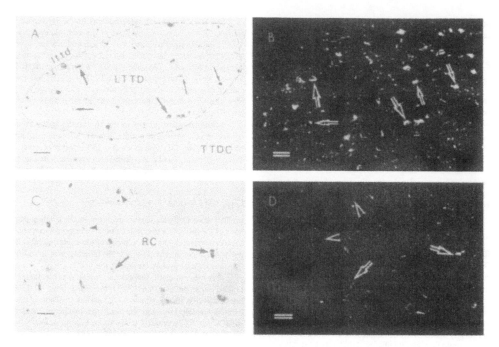

FIGURE 1. Brightfield (left) and darkfield (right) photomicrographs of the LTTD (A, B) and the RC (C, D), to show the cytoarchitecture and labeling patterns under basal conditions. Dorsal is upper and medial is right. (A, B) Large and small cells (large and small arrows, respectively) in the LTTD (within dashed line) are heavily labeled. Some glial cells (double arrow) are also labeled. However, the TTDC is not labeled. (C, D) The labeled cells (arrows) in the RC are less dense than in the LTTD. Some cells (arrowhead) lack labeling. LTTD, nucleus descendens lateralis n. trigemini; RC, nucleus reticularis caloris; TTDC, nucleus caudalis n. trigemini; lttd, tractus descendens lateralis n. trigemini. Calibration bars: 40 μm.

DG under basal conditions. Not only light and large cells, but also dark and small cells, showed labeling. Silver grains were deposited over the cytoplasm or at the outer boundary or at one end of the cells, which may have been the result of some movement or diffusion of 2-DG during histological treatment (Fig. 6). The labeling did not extend to any stem axons, and this was different from the multidendrites of motoneurons in the MT (Fig. 2). We did not find labeling in satellite cells of trigeminal ganglia.

We analyzed quantitatively the labeled cells of the ophthalmic ganglion as an example. Among 942 labeled cells ($n = 3$), 38.85% were light and 61.15% were dark cells. The percentage of labeled cells tended to increase ipsilaterally after unilateral IR stimulation. However, no significant difference ($p > 0.05$) was seen in means or population

distribution between the two sides in the three IR-stimulated animals.

DISCUSSION

The data presented here were obtained by following the high-resolution [³H]2-DG protocol of Durham et al. (1981) and McCasland and Woolsey (1988a,b). Undoubtedly, loss of much radioactivity is a shortcoming of this method, and presumably it occurred in our materials too (data from scintillation counting of washout solution are not shown). In our autoradiograms, the silver grains were mostly localized in somata and in their proximal processes in some places, but usually the nucleus was devoid of labeling (Figs. 2, 4). These cellular labelings were similar to those of other cell types (Durham et al., 1981; Kai-Kai and Pentreath, 1981;

FIGURE 2. Darkfield photomicrograph of the MT (lower part) and MTD (upper part), to show labeling patterns under basal conditions. The cells and proximal processes are filled with silver grains, but their nuclei are not. Arrows indicate labeled linear elements, probably neurites. Calibration bar: 40 μm.

Pentreath et al., 1982; Witkovsky and Yang, 1982; Durham and Woolsey, 1985; McCasland and Woolsey, 1988a). Silver grains in the neurons were attributable to 2-DG uptake and its metabolic products (Hammer and Herkenham, 1984).

[³H]2-DG Labeled Cells in Some Trigeminal and IR Nuclei under Basal Conditions

We were quite surprised at finding 2-DG-labeled cells in the LTTD and the RC after complete removal of the pit membrane(s) (Figs. 1, 2, 3). Because both of these are considered sensory nuclei of the IR system, we expected active discharge or 2-DG labeling to disappear after the removal of sensory input, as in reports from Auker et al. (1983) and Gruberg et al. (1984). The discrepancy between our results and theirs may be attributable mainly to the

different methodology. Light labeling is easily overlooked in the low resolution of [¹⁴C]2-DG autoradiography (Lippe et al., 1980; Sejnowski et al., 1980).

These unexpected labelings may indicate a high metabolic level. Our results showed many labeled neurons in the LTTD, the RC, the MRT, and the MT and MTD; this appeared to be consistent with the finding that these nuclei contain high succinate dehydrogenase (SDH) activity in snakes of another species, *Agkistrodon blomhoffi* (Kusunoki et al., 1987). High activity of SDH, probably together with other oxidative enzymes, may account for much 2-DG uptake, since the level of oxidative enzymes such as SDH and cytochrome oxidase has been suggested to reflect the level of glucose utilization (Marshall et al., 1981; Brown and Brunjes, 1990; Jacquin et al., 1993). In addition, the basal labeling may reflect an increased metabolic activity that maintains a level of spontaneous activity and a sensitivity to IR stimulus, since frequent and irregular spontaneous discharge has been electrophysiologically demonstrated in the LTTD (Terashima and Goris, 1977). Gruberg et al. (1984) recorded low-level discharges from the LTTD, but no response to IR stimulus, when the trigeminal nerve branches were cut. These were probably injury discharges (Govrin-Lippmann and Devor, 1978; Papir-Kricheli and Devor, 1988). A disinhibitory effect is also conceivable after removal of the pit organ. Neuronal activity independent of the pit receptor output may have occurred when the pit was removed and may have contributed to the basal labelings we found. Furthermore, it is necessary to consider whether ketamine as an anesthetic in our experiment could have produced elevated labeling regardless of stimulation, since ketamine increases spontaneous discharges in many brain regions (Warenycia and McKenzie, 1984). Our observations

TABLE 1. Percentages of Labeled Cells in Four Nuclei for Five Preparations under Basal Conditions

Preparation no.	LTTD		RC	MRT	MT and MTD
	Large cells	Small cells			
1	49.23	77.41	31.06	91.98	89.56
2	51.62	62.86	34.97	88.37	94.66
3	40.53	60.13	55.36	90.31	69.45
4	64.91	73.22	65.63	90.81	93.66
	68.30	74.67	67.71	91.90	93.64
5	32.52	68.89	38.14	89.44	89.92
	31.07	70.60	35.25	89.39	91.14
Mean	48.31	69.68[a]	46.87	90.31[b]	88.86[b]

Note. Data are from one side of the brain in preparations 1, 2, and 3, and from both sides in preparations 4 and 5.

[a] Significantly different from large cells in the LTTD and cells in the RC, $p < 0.05$.

[b] Significantly different from large and small cells in the LTTD and cells in the RC, $p < 0.05$.

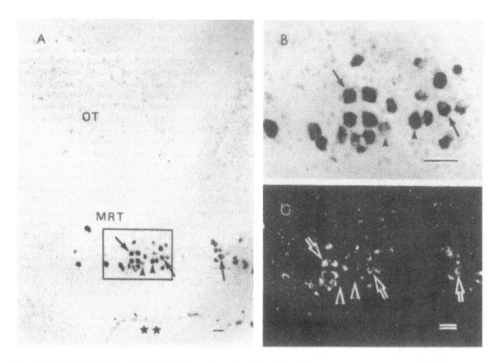

FIGURE 3. Brightfield (A, B) and darkfield (C) photomicrographs of the OT and the MRT, to show the cytoarchitecture and labeling patterns under basal conditions. (A, C) The MRT is labeled heavily and bilaterally (right is to the right). (B) A high magnification of the boxed area in A. A number of cells are labeled (arrows), but a few cells (arrowheads) are unlabeled, though they can be seen with dense Nissl counterstain in A and B. (A) The middle laminae are not labeled on either side of the OT. Stars indicate the ventricle, where labeling is absent. MRT, nucleus mesencephalicus n. trigemini; OT, optic tectum. Calibration bars: 40 μm.

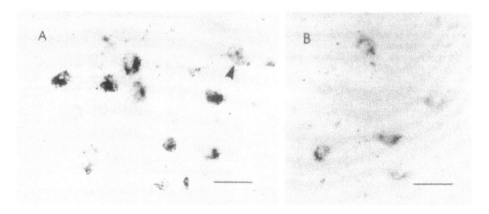

FIGURE 4. High-magnification photomicrographs of the LTTD (A) and the RC (B), to show labeled neurons after IR stimulation. The labeling density of neurons in the LTTD is usually heavier than in the RC. The arrowhead indicates an unlabeled large cell. Calibration bars: 40 μm.

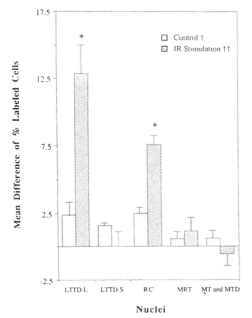

FIGURE 5. Mean differences in percentages of labeled cells between the two sides of the brain in four nuclei. †Both pits were removed as controls (n = 2). ††IR stimulation experiments (n = 3). LTTD-L and LTTD-S, large and small cells in the LTTD, respectively. Thin vertical bars indicate standard errors. *p < 0.05.

of various intervals between ketamine use and 2-DG injection suggest that it had no effect on our experiments.

IR-Evoked [³H]2-DG Labeling

The percentage of labeled cells increased when the ipsilateral pit was stimulated—by 12.84% in the LTTD (larger cells) and by 7.55% in the RC, as compared to the contralateral basal level. This indicates that about 12.84% of large cells in the LTTD and 7.55% of total cells in the RC were activated by the IR stimulation.

We found two labeling features in the LTTD. First, the percentage of large cells labeled during stimulation showed a small increase: 59.97% of the total number of large cells were labeled in the IR-stimulated side, which was only 1.27 times the number of the cells labeled under basal conditions (47.13%). McCasland and Woolsey (1988b) found that the number of neurons in rat C3 barrel column labeled with 2-DG when all whiskers were stimulated was about 10 times the number of cells labeled when all whiskers were clipped. Our results show a comparatively slight increase in the num-

ber of cells labeled by stimulation, indicating only a small increase in glucose consumption during IR reception. Since Fox et al. (1988) demonstrated a 50% increase of regional blood flow in human visual cortex during visual stimulation, regional blood flow in the LTTD may increase during pit stimulation. The rich capillary network in this nucleus would meet this functional demand (Meszler et al., 1981) although at this moment we are unable to say whether the energy is produced mainly by glucose oxidation or not. Second, numerous small cells in the LTTD were labeled (Fig. 1, Table 1), and this was not changed by IR stimulation (Fig. 5). These cells appear to be interneurons, whose axon terminals contain pleomorphic vesicles and much glycogen. The terminals form axoaxonic synapses with the afferent terminals to modify the excitatory relay (Meszler, 1981; Meszler et al., 1981). The number of such axoaxonic synapses is relatively high, accounting for one-third of the synapses in the nucleus. Our results suggest that 2-DG uptake in these small neurons, probably inhibitory interneurons, may be relatively constant.

Although there is evidence that glial cells incorporate glucose into glycogen and affect neuronal activity (Kai-Kai and Pentreath, 1981; Tsacopoulos et al., 1988; Swanson, 1992; Swanson et al., 1992), our materials only displayed a small amount of labeling on glial cells, which appeared not to be changed with IR stimulation.

IR stimulation did not produce [³H]2-DG labeling in the OT in our snakes. This finding is consistent with that of Auker et al. (1983), but contradictory to that of Gruberg et al. (1984). The 2-DG labeling seems to be progressively reduced from lower to higher central relays. Our results, like those of Auker and Gruberg, showed denser labeling in the LTTD, lighter labeling in the RC, and absent or very light labeling in the OT (Figs. 1, 4), just as Webster er al. (1984) found selective 2-DG labeling in the monkey auditory system. The reason is unknown. The brainstem trigeminal nuclei are the first relay stations, and they are likely to require more energy than higher central structures when there are rapid and prolonged changes in primary afferent input (Yip et al., 1987). At higher levels, only a small number of generated spikes can be detected (Auker et al., 1983). This seems to be the reason why higher central relays lack labeling.

Comparison between the Trigeminal Ganglia and the MRT

Although both the trigeminal ganglia and the MRT contain first-order neurons, they exhibited different labeling patterns: There were more labeled neurons in the MRT than in the trigeminal ganglia (Figs. 3, 6).

The [³H]2-DG labeling in the trigeminal ganglia was seen in small and large as well as dark and light cells, and in the mandibular divisions that do not innervate the pit organs; moreover, there was no change in population distribution of labeled cells in the ipsilateral ophthalmic ganglion after IR stimulation. Therefore, 2-DG labeled ganglion cells seem not to be exclusively related to IR stimulation.

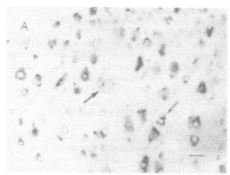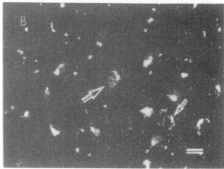

FIGURE 6. Brightfield (A) and darkfield (B) photomicrographs of the maxillomandibular ganglia, to show the cytoarchitecture and labeling patterns under basal conditions. Large and small arrows indicate a labeled large light cell and a small dark cell, respectively. The silver grains are on cell bodies without extending to any axons. Satellite cells lack labeling. Calibration bars: 40 μm.

In conclusion, the present study showed more details of [³H]2-DG labeling patterns in the snake trigeminal system than previous research has revealed. Under basal conditions, some trigeminal nuclei manifested labeling, in which the LTTD exhibited high glucose utilization but the common trigeminal sensory nuclei did not. This labeling seems to be caused by neuronal oxidative enzyme activity and spontaneous activity of the cells. By quantitative analysis of neuron counts, we found that the percentage of labeled cells increased slightly in the LTTD large cells and in the RC after IR stimulation, suggesting only a small increase of glucose consumption. On the other hand, 2-DG uptake in the small neurons of the LTTD, which are probably inhibitory interneurons, was relatively constant.

ACKNOWLEDGMENTS

We wish to thank Prof. T. Kusunoki for helping with anatomical identification and Prof. V. Mizuhira for teaching us autoradiographic techniques.

REFERENCES

AUKER, C. R., R. M. MESZLER, and D. O. CARPENTER (1983) Apparent discrepancy between single-unit activity and [¹⁴C]deoxyglucose labeling in optic tectum of the rattlesnake. J. Neurophysiol. 49: 1504–1516.

BROWN, J. L., and P. C. BRUNJES (1990) Development of the anterior olfactory nucleus in normal and unilaterally odor deprived rats. J. Comp. Neurol. 301: 15–22.

DE COCK BUNING, T., S. TERASHIMA, and R. C. GORIS (1981) Crotaline pit organs analyzed as warm receptors. Cell. Mol. Neurobiol. 1: 69–85.

DURHAM, D., and T. A. WOOLSEY (1985) Functional organization in cortical barrels of normal and vibrissae-damaged mice: A (³H) 2-deoxyglucose study. J. Comp. Neurol. 235: 97–110.

DURHAM, D., T. A. WOOLSEY, and L. KRUGER (1981) Cellular localization of 2-[³H]deoxy-D-glucose from paraffin-embedded brains. J. Neurosci. 1: 519–526.

FOX, P. T., M. E. RAICHLE, M. A. MINTUN, and C. DENCE (1988) Nonoxidative glucose consumption during focal physiologic neural activity. Science 241: 462–464.

GRUBERG, E. R., E. A. NEWMAN, and P. H. HARTLINE (1984) 2-Deoxyglucose labeling of the infrared sensory system in the rattlesnake, Crotalus viridis. J. Comp. Neurol. 229: 321–328.

GOVRIN-LIPPMANN, R., and M. DEVOR (1978) Ongoing activity in severed nerves: Source and variation with time. Brain Res. 159: 406–410.

HAMMER, R. P., JR., and M. HERKENHAM (1984) Tritiated 2-deoxy-D-glucose: A high-resolution marker for autoradiographic localization of brain metabolism. J. Comp. Neurol. 222: 128–139.

JACQUIN, M. F., J. S. McCASLAND, T. A. HENDERSON, R. W. RHOADES, and T. A. WOOLSEY (1993) 2-DG uptake patterns related to single vibrissae during exploratory behaviors in the hamster trigeminal system. J. Comp. Neurol. 332: 38–58.

KAI-KAI, M. A., and V. W. PENTREATH (1981) High resolution analysis of [³H]-deoxyglucose incorporation into neurons and glial cells in invertebrate ganglia: Histological processing of nervous tissue for selective marking of glycogen. J. Neurocytol. 10: 693–708.

KISHIDA, R., S. TERASHIMA, R. C. GORIS, and T. KUSUNOKI (1982) Infrared sensory neurons in the trigeminal ganglia of crotaline snakes: Transganglionic HRP transport. Brain Res. 241: 3–10.

KONIGSMARK, B. W. (1970) Methods for the counting of neurons. In Contemporary Research Methods in Neuroanatomy, W. J. H. Nauta and S. O. E. Ebbesson, eds., pp. 315–340, Springer-Verlag, Berlin.

KUSUNOKI, T., R. KISHIDA, T. KADOTA, and R. C. GORIS (1987) Chemoarchitectonics of the brainstem in infrared sensitive and nonsensitive snakes. J. Hirnforsch. 28: 27–43.

LIPPE, W. R., O. STEWARD, and E. W. RUBEL (1980) The effect of unilateral basilar papilla removal upon nuclei laminaris and magnocellularis of the chick examined with [³H]2-deoxy-D-glucose autoradiography. Brain Res. 196: 43–58.

MARSHALL, J. F., J. W. CRITCHFIELD, and M. R. KOZLOWSKI (1981) Altered succinate dehydrogenase activity of basal ganglia following damage to mesotelencephalic dopaminergic projection. Brain Res. 212: 367–377.

McCasland, J. S., and T. A. Woolsey (1988a) New high-resolution 2-deoxyglucose method featuring double labeling and automated data collection. J. Comp. Neurol. *278*: 543–554.

McCasland, J. S., and T. A. Woolsey (1988b) High-resolution 2-deoxyglucose mapping of functional cortical columns in mouse barrel cortex. J. Comp. Neurol. *278*: 555–569.

McLean, I. W., and P. K. Nakane (1974) Periodate–lysine–paraformaldehyde fixative: A new fixative for immunoelectron microscopy. J. Histochem. Cytochem. *22*: 1077–1083.

Meszler, R. M. (1983) Fine structure and organization of the infrared receptor relay: Lateral descending nucleus of V in boidae and nucleus reticularis caloris in the rattlesnake. J. Comp. Neurol. *220*: 299–309.

Meszler, R. M., C. R. Auker, and D. O. Carpenter (1981) Fine structure and organization of the infrared receptor relay, the lateral descending nucleus of the trigeminal nerve in pit vipers. J. Comp. Neurol. *196*: 571–584.

Nelson, T., E. E. Kaufman, and L. Sokoloff (1984) 2-Deoxyglucose incorporation into rat brain glycogen during measurement of local cerebral glucose utilization by the 2-deoxyglucose method. J. Neurochem. *43*: 949–956.

Newman, E. A., E. R. Gruberg, and P. H. Hartline (1980) The infrared trigemino-tectal pathway in the rattlesnake and in the python. J. Comp. Neurol. *191*: 465–477.

Northcutt, R. G. (1984) Anatomical organization of the optic tectum in reptiles. In *Comparative Neurology of the Optic Tectum*, H. Vanegas, ed., pp. 547–600, Plenum Press, New York.

Nudo, R. J., and R. B. Masterton (1986) Stimulation-induced [¹⁴C]2-deoxyglucose labeling of synaptic activity in the central auditory system. J. Comp. Neurol. *245*: 553–565.

Papir-Kricheli, D., and M. Devor (1988) Abnormal impulse discharge in primary afferent axons injured in the peripheral versus the central nervous system. Somatosens. Mot. Res. *6*: 63–77.

Pentreath, V. W., L. H. Seal, and M. A. Kai-Kai (1982) Incorporation of [³H]2-deoxyglucose into glycogen in nervous tissues. Neuroscience *7*: 759–767.

Pilgrim, C., and H. J. Wagner (1981) Improving the resolution of the 2-deoxy-D-glucose method. J. Histochem. Cytochem. *29*: 190–194.

Sejnowski, T. J., S. C. Reingold, D. B. Kelley, and A. Gelperin (1980) Localization of [³H]-2-deoxyglucose in single molluscan neurons. Nature *287*: 449–451.

Sharp, F. R., J. S. Kauer, and G. M. Shepherd (1975) Local sites of activity-related glucose metabolism in rat olfactory bulb during olfactory stimulation. Brain Res. *98*: 596–600.

Sharp, F. R., S. M. Sagar, and R. A. Swanson (1993) Metabolic mapping with cellular resolution: c-*fos* vs. 2-deoxyglucose. Crit. Rev. Neurobiol. *7*: 205–228.

Sokoloff, L., M. Reivich, C. Kennedy, M. H. Des Rosiers, C. S. Patlak, K. D. Pettigrew, O. Sakurada, and M. Shinohara (1977) The [¹⁴C]deoxyglucose method for the measurement of local cerebral glucose utilization: Theory, procedure, and normal values in the conscious and anesthetized albino rat. J. Neurochem. *28*: 897–916.

Stanford, L. R., D. M. Schroeder, and P. H. Hartline (1981) The ascending projection of the nucleus of the lateral descending trigeminal tract: A nucleus in the infrared system of the rattlesnake, *Crotalus viridis*. J. Comp. Neurol. *201*: 161–173.

Swanson, R. A. (1992) Physiologic coupling of glial glycogen metabolism to neuronal activity in brain. Can. J. Physiol. Pharmacol. *70* (Suppl.): 138–144.

Swanson, R. A., M. M. Morton, S. M. Sagar, and F. R. Sharp (1992) Sensory stimulation induces local cerebral glycogenolysis: Demonstration by autoradiography. Neuroscience *51*: 451–461.

Terashima, S., and R. C. Goris (1977) Infrared bulbar units in crotaline snake. Proc. Japan. Acad. B *53*: 292–296.

Terashima, S., and P.-J. Jiang (1994) 2-Deoxyglucose labeling of the trigeminal sensory system. Japan. J. Physiol. *44* (Suppl. 1): 204.

Tsacopoulos, M., V. Evequoz-Mercier, P. Perrottet, and E. Buchner (1988) Honeybee retinal glial cells transform glucose and supply the neurons with metabolic substrate. Proc. Natl. Acad. Sci. USA *85*: 8727–8731.

Warenycia, M. W., and G. M. McKenzie (1984) Responses of striatal neurons to anesthetics and analgesics in freely moving rats. Gen. Pharmacol. *15*: 517–522.

Webster, W. R., J. Serviere, D. Crewther, and S. Crewther (1984) Iso-frequency 2-DG contours in the inferior colliculus of the awake monkey. Exp. Brain Res. *56*: 425–437.

Witkovsky, P., and C.-Y. Yang (1982) Uptake and localization of ³H-2-deoxy-D-glucose by retinal photoreceptors. J. Comp. Neurol. *204*: 105–116.

Wolfe, D. E., and J. G. Nicholls (1967) Uptake of radioactive glucose and its conversion to glycogen by neurons and glial cells in the leech central nervous system. J. Neurophysiol. *30*: 1593–1609.

Yip, V. S., W.-P. Zhang, T. A. Woolsey, and O. H. Lowry (1987) Quantitative histochemical and microchemical changes in the adult mouse central nervous system after section of the infraorbital and optic nerves. Brain Res. *406*: 157–170.

Temperature-Induced Changes in the Number of Vesicles in the Free Nerve Endings of Temperature Neurons of the Snake

Shin-ichi Terashima,*[,1] Peng-Jia Jiang,* Vinci Mizuhira,[†] Hiroshi Hasegawa,[†] and Mitsuru Notoya[†]

*Department of Physiology, University of the Ryukyus School of Medicine, Nishihara-cho, Okinawa 903-01, Japan; [†]Shionogi & Co. Ltd., Research Laboratories, Kanzakigawa Laboratory, Osaka 561, Japan

Abstract   By observing ultrastructural changes under the electron microscope, we illustrated exocytosis and recycling of vesicles in the infrared receptor, a kind of free nerve ending in the pit organ of the crotaline snake, *Trimeresurus flavoviridis*. While maintaining the snake pit organs at stable temperatures of 15°C, 25°C, and 30°C, we fixed them by perfusion and then processed them for transmission electron microscopy. The largest number of clear and coated vesicles appeared in the terminals at the lowest temperature. The perimeter and area of a terminal were enlarged at 30°C, and "opening waves" on the plasma were prominently found at the highest temperature. We also observed coated vesicles that budded from the plasma membrane in the terminals. The configuration of mitochondria in the terminals was quantitatively different between lower and higher temperatures. The data suggest that exocytosis and endocytosis in these terminals operate in a manner similar to that observed in other cell types.

Key words   exocytosis, free nerve ending, temperature neuron, pit organ, snake, vesicle, mitochondria

Vesicles in the nervous system are found not only in synaptic terminals, but also in free axonal endings of infrared (IR) or thermal receptors of the crotaline snake (Terashima et al., 1970; Hirosawa, 1980). We have been puzzled about why the vesicles exist in such nonsynaptic receptor endings, which are also rich in mitochondria.

The function of the pit organs of snakes has been examined electrophysiologically by several groups of researchers (Bullock and Diecke, 1956; Goris and Nomoto, 1967; Warren and Proske, 1968; Harris and Gamow, 1971; DeSalvo and Hartline, 1978; Terashima and Goris, 1983). Trigeminal nerve or ganglion cell discharges are very sensitive to heat applied transiently to the pit organ. Responses adapt at steady temperatures (Bullock and Diecke, 1956; Terashima and Goris, 1979; de Cock Buning et al., 1981). In early experiments, Terashima et al. (1968) demonstrated generator potential in the pit membrane with IR stimulation. However, it is still a mystery how IR energy is transduced into a generator potential.

Synaptic terminals generate postsynaptic potentials by releasing transmitters from vesicles, whose coupled but independent exocytosis–recycling process is well known (Heuser and Reese, 1973; Model et al., 1975; Rose et al.,

1978; Schaeffer and Raviola, 1978; Weldon et al., 1990). Exocytosis and endocytosis are also involved in secretion of secretory glands (Nagasawa et al., 1970) as well as in other functions in taste buds (Farbman and Hellekant, 1989), cochlear hair cells (Hackney et al., 1993), micropores of *Toxoplasma gondii* (Nichols et al., 1994), and so on. In the present study, we used ultrastructure for the first time to examine the events at sensory nerve endings of temperature neurons of the snake and to relate our observations to temperature changes.

A preliminary report of this research has been presented elsewhere (Terashima and Jiang, 1993).

MATERIALS AND METHODS

Animal Treatment

Eight crotaline snakes (*Trimeresurus flavoviridis*), which possess a pair of pit organs on the face, were studied. After administration of halothane anesthesia and pancuronium immobilization (2 mg/kg i.m.), each snake was perfused through the carotids of both sides with two micropumps at room temperature (ca. 25°C) with heparinized saline for 10 to 30 min, followed by 200 ml of a fixative composed of 2% paraformaldehyde, 2% glutaraldehyde, and 0.1% tannic

1. To whom all correspondence should be addressed.

acid in 0.1 M cacodylate buffer (pH 7.4) for about 30 min or longer (Mizuhira and Futaesaku, 1972; Mizuhira et al., 1990). During this period, the head of the snake was kept in a beaker filled with water at either 15°C (n = 3), 25°C (n = 1), or 30°C (n = 2). Another two animals were used for a recovery experiment; these snakes were heated to 30°C for 2 min and then cooled to 15°C for 10 min.

To follow vesicle recycling, we studied uptake of horseradish peroxidase (HRP) into vesicles in five other snakes. After intracardiac injection of HRP (Sigma, type VI, 87 mg in 1 ml saline for one snake), these animals were kept for 2 min at 30°C, and then for 10 min at 15°C.

Electron Microscopy

The pit membranes were carefully dissected immediately after perfusion and cut into pieces (2 × 5 mm²). The specimens were then immersed in fresh fixative solution at 4°C overnight. After three 20-min rinses with 0.1 M cacodylate buffer, the specimens were postfixed in 1% osmium tetroxide in 0.1 M cacodylate buffer for 2 hr, dehydrated through a graded series of ethanol solutions followed by propylene oxide, and embedded in Epon 812 epoxy resin by conventional methods (Sevéus and Johannessen, 1978).

The specimen blocks were sectioned on an Ultracut (Reichert-Jung, Vienna). Semithin sections were cut with a glass knife and stained with 0.5% toluidine blue. Ultrathin sections were cut with a diamond knife (Nanotome, Akashi Beam Tech.) at 50–70 nm and mounted on formvar-covered grid meshes. After double staining with uranyl acetate and lead citrate on the grid, the sections were observed under an electron microscope (JEM-2000 EX, JEOL Ltd.).

We collected four blocks per animals, and three micrographs of different parts of each block, which were selected randomly. The nerve endings were photographed at × 10,000 and × 20,000 and enlarged to 15 × 20 cm² (actual area = 7.68 μm²) for measurement. Parameters (area, perimeter, and diameter of structures) were measured on the prints of 12 pictures per animal with a Nikon three-dimension analyzer system (Cosmozone 98, Nippon Kogaku K.K.). Since 25°C was an intermediate condition, and there was only a single 25°C case, we concentrated our statistical analysis by comparing 15°C and 30°C animals. One-way analysis of variance (ANOVA; StatView 4.0) software was used with a Macintosh computer. p < 0.05 was taken as a statistically significant level of difference.

RESULTS

Ultrastructure of the Pit Membrane

Our observations on the structure of the pit membrane were similar to those of earlier reports (Bleichmar and De Robertis, 1962; Terashima et al., 1970; Hirosawa, 1980). This thin membrane consists of an inner and outer epithelial layer, and an intervening nerve layer. In the nerve layer, free nerve endings supported by Schwann cells, and myelinated and nonmyelinated fibers, were observed. The free nerve endings had irregular profiles and various sizes, containing mitochondria, clear vesicles, and vacuoles (Figs. 1, 2, 3); we did not find endoplasmic reticulum or Golgi complex.

Mitochondria

Mitochondria were tightly packed, and some appeared to be dividing. We observed an unusual structure between abutting mitochondria; this appeared as a series of regular septa in tangential views (Fig. 4), similar in structure to a septate desmosome (Wood, 1959, *Hydra*; Gilula et al., 1970, mussel). The mitochondria appeared to be held in close contact with one another by these structures.

Vesicles and Coated Vesicles

Vesicles have been reported previously (Terashima et al., 1970; Hirosawa, 1980), whereas coated vesicles were observed here for the first time. In the free endings, the vesicles were clear and round, with a mean diameter of 42.88 ± 7.08 nm (n = 2114). Coated vesicles had radiating spokes, resembling the appearance of clathrin coats described in other cell types (see Pley and Patham, 1993). They mingled with clear vesicles of similar or smaller diameter (Figs. 1, 8). The coated vesicles budded from plasma membrane (Fig. 5) and formed cisternae (Fig. 6).

Attempts to observe the uptake of tracer into vesicles following intracardiac injection of HRP were unsuccessful. No electron-dense reaction products were observed either in vesicles or in perineuronal spaces, probably because the terminals were inaccessible to the tracer by conventional techniques.

Free Nerve Endings at Extreme Temperature Conditions

MORPHOLOGY

The mean diameters of the vesicles did not differ with temperature and were 42.92 nm (± 7.18 *SD*, n = 1450), 42.98 nm (± 6.47 *SD*, n = 268), and 42.79 nm (± 7.25 *SD*, n = 396) at 15°C, 25°C, and 30°C, respectively. We found more abundant clear vesicles in the free nerve endings at 15°C (Fig. 1) than at 25° and 30°C (Figs. 2, 3). The vesicles were located principally near the plasma membrane.

On the other hand, after exposure of a snake to heat (30°C), there were only a few vesicles that were scattered sporadically in the terminals. At this temperature, some "omega" figures and "waves" were prominent on the terminal membrane (Fig. 7). The perimeter and area of each terminal at 30°C were larger than at lower temperatures (Figs. 1, 3).

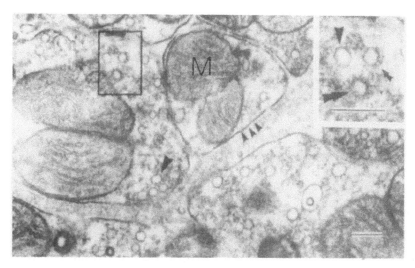

FIGURE 1. Electron micrograph of free nerve endings at 15°C. Vesicles are abundant (arrowheads). Condensed mitochondria (M) fill the terminals, which have smooth plasma membranes (three small arrowheads). Some coated vesicles (double arrowheads) mingle with clear vesicles; their radiating spokes are evident at the upper right corner at high magnification in the boxed area. A small double arrowhead indicates a half-coated vesicles, and a large double arrowhead a fully coated one. Calibration bars in this and subsequent figures: 200 nm.

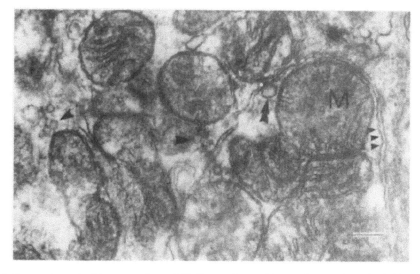

FIGURE 2. Electron micrograph of free nerve endings at 25°C. There are fewer vesicles (arrowheads) and vacuoles (double arrowhead), larger mitochondria (M), and wavy plasma membrane (three small arrowheads).

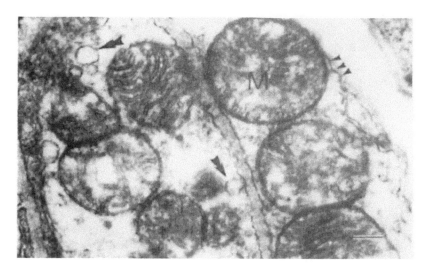

FIGURE 3. Electron micrograph of free nerve endings at 30°C. Mitochondria (M) are rounder, are swollen, and possess lighter matrices, with vacuoles (double arrowheads) and wavy plasma membrane (three small arrowheads).

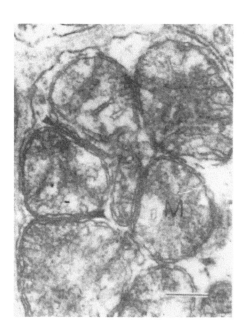

FIGURE 4. Electron micrograph of structures resembling septate desmosomes. Arrowheads indicate the novel structures between the mitochondria (M).

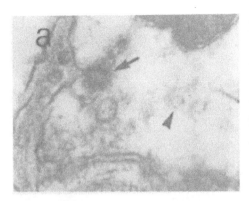
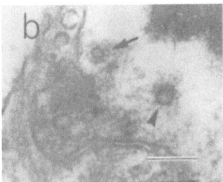

FIGURE 5. Electron micrograph of coated vesicles budding from the plasma membrane at 15°C in two serial sections. (a) A coated pit (arrow) buds from the plasma membrane. Its narrow neck and opening mouth can barely be seen. (b) This coated vesicle (arrow) appears within the nerve ending. Arrowheads indicate another coated vesicle nearby.

The configuration of mitochondria was changed at 30°C as compared to 15°C, being enlarged overall with lighter matrices and wider cristae at higher temperatures (Figs. 1, 3).

In contrast to these changes in the terminal, Schwann cell somata had no noticeable changes at three different temperatures.

QUANTITATIVE ANALYSIS

The total number of vesicles at 15°C was 2.4- and 1.8-fold greater than the numbers of 30°C ($p < 0.05$) and 25°C, respectively (Table 1). The number of coated vesicles was greatest at 15°C. The mean perimeter and area per terminal at 30°C were significantly larger than at 15°C ($p < 0.05$, Table 2). The number of mitochondria was higher at 15°C than at 30°C ($p < 0.05$, Table 3). The values for mitochondrial size, short–long diameter ratio, and mean area confirmed their overall enlargement at 30°C.

Free Nerve Endings in the Recovery Condition

The structural changes were reversed when snakes were cooled (15°C) for 10 min after 2 min of heating conditions (30°C) (Fig. 8). The number of vesicles, the area per terminal, and the number of vacuoles returned to a state similar

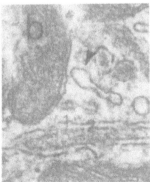

FIGURE 6. Electron micrograph of coated vesicles forming a cisterna at 15°C in three serial sections. A coated vesicle (1 in a) fuses with another (2 in b) to make a cisterna (arrow in c). Arrowheads indicate another fused cisterna nearby.

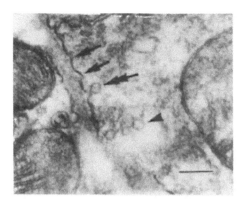

FIGURE 7. Electron micrograph of exocytosis in a free nerve ending at 30°C. Note the opening or omega figure (double arrow) at the plasma membrane from a smooth vesicle. Several vesicles (arrowhead) and some "opening wave" shapes (arrows) are nearby.

to that at 15°C. The size of mitochondria was reduced, and the matrices became denser (Fig. 8).

DISCUSSION

Coated Vesicles in Free Nerve Endings

We observed a number of vesicles and coated vesicles in the nerve terminals (Fig. 1). Clathrin-like coats were found with the coated pits and vesicles, which budded from the cytoplasmic membrane (Fig. 5) and fused with one another to form cisternae (Fig. 6), suggesting that vesicles are in continuity with the plasma membrane.

Studies of coated pits or vesicles in other non-neuronal cells have shown that there are two classes of coated vesicles (cf. Nichols et al., 1994). Pinocytic vesicles are large in size (about 150 nm) and usually have clathrin coats, whereas those originating in the Golgi region are smaller and have coats consisting of different proteins. Since there is no Golgi complex in the terminal, because it is remote from the cell body, these results suggest that the coated pits and vesicles are probably clathrin-coated at the terminal surface, perhaps via receptor-mediated endocytosis.

Temperature-Dependent Exocytosis and Endocytosis

The present results showed that the number of vesicles in the terminals decreased or increased reversibly with temperatures (Figs. 1, 2, 3, 8). Concomitant with this change, the perimeter and area of the terminal enlarged or contracted. In our specimens we found "omega" figures on the plasma membrane of the terminal, although it had no synapse-like

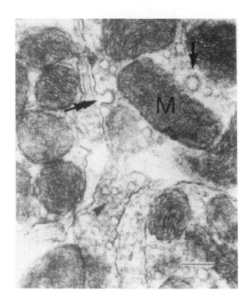

FIGURE 8. Electron micrograph of a free nerve ending 10 min after recovery. Many vesicles (arrowhead) have been recovered. Mitochondria (M) are reduced in size, with condensed cristae and dense matrices. A coated vacuole (arrow) and an opening coated pit (double arrow) are seen.

structure (Fig. 7). Our ultrastructural observations suggest that exocytosis and endocytosis of vesicles occur at the free nerve endings of temperature neurons of the snake, which are very sensitive to temperature changes.

It is still unknown why the exocytosis and endocytosis at this terminal are so strikingly temperature-dependent. However, this phenomenon coincides with electrophysiological evidence that the discharge rate of this receptor is strongly affected by temperature changes (Bullock and Diecke, 1956; Goris and Nomoto, 1967; Terashima and

TABLE 1. Quantitative Properties of Vesicles

	15°C ($n = 3$)	25°C ($n = 1$)	30°C ($n = 2$)
Total number of vesicles	483.33 ± 20.88	268	$198.00 \pm 4.00^*$
Coated vesicles	42 ± 3.06 (8.67%)	20 (7.46%)	$8.50 \pm 0.50^*$ (4.29%)

Note. The values are means $\pm$ SDs in 92.16 μm^2 (7.68 $\mu m^2 \times$ 12) area measured. n is number of animals. The values in parentheses are percentages of coated vesicles to total vesicles.

$^*p < 0.05$ (the 15°C vs. the 30°C group).

TABLE 2. Mean Perimeter and Area per Terminal at 15°C and 30°C

	15°C	30°C
Perimeter (μm)	4.39 ± 0.16	6.21 ± 0.17*
Area (μm²)	1.26 ± 0.08	2.32 ± 0.10*

Note. The values are means ± SDs. The data were measured on 150 profile terminals at 15°C and 30°C.
*p < 0.05.

Goris, 1983). Electrophysiological study proved that this organ is especially sensitive to temperature changes, but discharge rate at high static temperatures is not very high. It is possible that this results from depletion of vesicles in the terminals, as in synaptic terminals. If exocytosis is operative in the terminals, it may be relevant for interpreting the discharge properties from the stimulus–response curve of this receptor at steady temperatures, which is bell-shaped (de Cock Buning et al., 1981); the static discharge rate decreases at high (30°C) and at low (15°C) temperature, with a peak of 25°C. Although the discharge rate is low at both ends, the possible mechanisms may be different; at the low temperature the receptor may have reduced biological activity in exocytosis (Fig. 1), producing a low discharge rate, whereas at the high temperature depletion of vesicles (Fig. 3) may be limiting the receptor discharges.

We suggest that the exocytosis and endocytosis probably take place in the pit membrane, and that the vesicles and coated pits may constitute a pathway for extrusion of some cellular component or uptake of extracellular materials. However, we cannot determine with certainty whether or not this process is related to neurotransmitter release from vesicles as in synapse, since we have no evidence of any neurotransmitter in this terminal at this time. If substances are released from the vesicles, they could act upon the autoreceptors of the plasma membrane (cf. Chesselet, 1984), altering the ionic permeability for generating the receptor potential (Terashima et al., 1968). A similar mechanism is

TABLE 3. Quantitative Properties of Mitochondria

	15°C (n = 3)	25°C (n = 1)	30°C (n = 2)
Mean size (μm × μm)	0.30 × 0.50	0.39 × 0.56	0.46 × 0.60
Mean area (μm²)	0.13 ± 0.02	0.18	0.23 ± 0.03*
Ratio (short–long diameter)	0.63 ± 0.01	0.72	0.80 ± 0.06*
Total number (per 92.16 μm²)	239.67 ± 5.84	195	163 ± 2.00*

Note. The values are means ± SDs except for mean size. *n* is number of animals.
*p < 0.05 (the 15°C vs. the 30°C group).

conceivable for other thermoreceptors that have similar nerve endings (cf. Hensel, 1981).

Morphological Changes of Mitochondria

Meszler (1970) first reported mitochonrdial changes in the nerve terminals of temperture neurons after stimulation, in snakes of the genera *Agkistrodon*, *Python*, and *Corallus*. He attributed such changes to energy drain for the sodium pump. In our experiments, the swollen mitochondria appeared simultaneously with vesicle decrease at 30°C (Table 3 and Fig. 3) and recovered with cooling of the pit membrane (Fig. 7). Similar changes have been reported at synapses, due to the release of neurotransmitters (Jones and Kwanbunbumpen, 1970; Heuser and Reese, 1973; Pysh and Wiley, 1974; Rose et al., 1978; Basbaum and Heuser, 1979). High numbers of mitochondria and their configurational changes suggest that high energy demand occurs with activity in the pit membrane.

In conclusion, the present results are consistent with vesicle exocytosis and endocytosis in the free nerve endings of snake temperature neurons. It remains uncertain whether these processes are independent from that of transmitter release.

ACKNOWLEDGMENTS

We wish to thank Dr. L. Kruger and Dr. T. A. Woolsey for critical reading and comments on the manuscript, and Dr. R. C. Goris for editing the English.

REFERENCES

BASBAUM, C. B., and J. E. HEUSER (1979) Morphological studies of stimulated adrenergic axon varicosities in the mouse vas deferens. J. Cell Biol. *80*: 310–325.
BLEICHMAR, H., and E. DE ROBERTIS (1962) Submicroscopic morphology of the infrared receptor of pit vipers. Z. Zellforsch. Mikroscop. Anat. *56*: 748–761.
BULLOCK, T. H., and F. D. J. DIECKE (1956) Properties of an infrared receptor. J. Physiol. (Lond.) *134*: 47–87.
CHESSELET, M.-F. (1984) Presynaptic regulation of neurotransmitter release in the brain: Facts and hypothesis. Neuroscience *12*: 347–375.
DE COCK BUNING, TJ., S. TERASHIMA, and R. C. GORIS (1981) Crotaline pit organs analyzed as warm receptors. Cell. Mol. Neurobiol. *1*: 69–85.
DESALVO, J. A., and P. H. HARTLINE (1978) Spatial properties of primary sensory neurons in Crotalidae. Brain Res. *142*: 338–342.
FARBMAN, A. I., and G. HELLEKANT (1989) Evidence for a novel mechanism of binding and release of stimuli in the primate taste bud. J. Neurosci. *9*: 3522–3528.
GILULA, N. B., D. BRANTON, and P. SATIR (1970) The septal junction: A structural basis for intercellular coupling. Proc. Natl. Acad. Sci. USA *67*: 213–220.
GORIS, R. C., and M. NOMOTO (1967) Infrared reception in Oriental crotaline snakes. Comp. Biochem. Physiol. *23*: 879–892.
HACKNEY, C. M., R. FETTIPLACE, and D. N. FURNESS (1993) The functional morphology of stereociliary bundles on turtle cochlear hair cells. Hearing Res. *69*: 163–175.

HARRIS, J. F., and R. I. GAMOW (1971) Snake infrared receptors: Thermal or photochemical mechanism? Science *172*: 1252–1253.

HENSEL, H. (1981) *Thermoreception and Temperature Regulation*, Academic Press, London.

HEUSER, J. E., and T. S. REESE (1973) Evidence for recycling of synaptic vesicle membrane during transmitter release at the frog neuromuscular junction. J. Cell Biol. *57*: 315–344.

HIROSAWA, K. (1980) Electron microscopic observations on the pit organ of a crotaline snake *Trimeresurus flavoviridis*. Arch. Histol. Japan. *43*: 65–77.

JONES, S. F., and S. KWANBUNBUMPEN (1970) The effects of nerve stimulation and hemicholinium on synaptic vesicles at the mammalian neuromuscular junction. J. Physiol. (Lond.) *207*: 31–50.

MESZLER, R. M. (1970) Correlation of ultrastructure and function. In *Biology of the Reptilia*, Vol. 2, *Morphology*, B. C. Gans and T. S. Parsons, eds., pp. 305–314, Academic Press, London.

MIZUHIRA, V., and Y. FUTAESAKU (1972) New fixation for biological membranes using tannic acids. Acta Histochem. Cytochem. *5*: 233–235.

MIZUHIRA, V., H. HASEGAWA, and M. NOTOYA (1990) Demonstration of membrane associated calcium ions of X-ray microanalysis after microwave fixation. J. Clin. Electron Microsc. *23*: 5–6.

MODEL, P. G., S. M. HIGHSTEIN, and M. V. L. BENNETT (1975) Depletion of vesicles and fatigue of transmission at a vertebrate central synapse. Brain Res. *98*: 209–228.

NAGASAWA, J., W. W. DOUGLAS, and A. SCHULZ (1970) Ultrastructural evidence of secretion by exocytosis and of "synaptic vesicle" formation in posterior pituitary glands. Nature *227*: 407–409.

NICHOLS, B. A., M. L. CHIAPINO, and C. E. N. PAVESIO (1994) Endocytosis at the micropore of *Toxoplasma gondii*. Parasitol. Res. *80*: 91–98.

PLEY, U., and P. PATHAM (1993) Clathrin: Its role in receptor-mediated vesicular transport and specialized function in neurons. Crit. Rev. Biochem. Mol. Biol. *28*: 431–464.

PYSH, J. J., and R. G. WILEY (1974) Synaptic vesicle depletion and recovery in cat sympathetic ganglia electrically stimulated *in vivo*: Evidence for transmitter secretion by exocytosis. J. Cell Biol. *60*: 365–374.

ROSE, S. J., G. D. PAPPAS, and M. E. KRIEBEL (1978) The fine structure of identified frog neuromuscular junctions in relation to synaptic activity. Brain Res. *144*: 213–239.

SCHAEFFER, S. F., and E. RAVIOLA (1978) Membrane recycling in the cone cell endings of the turtle retina. J. Cell Biol. *79*: 802–825.

SEVÉUS, L., and J. V. JOHANNESSEN (1978) Embedding, sectioning, and staining. In *Electron Microscopy in Human Medicine*, Vol. 1, J. V. Johannessen, ed., pp. 116–184, McGraw-Hill, New York.

TERASHIMA, S., and R. C. GORIS (1979) Receptive areas of primary infrared afferent neurons in crotaline snakes. Neuroscience *40*: 1137–1144.

TERASHIMA, S., and R. C. GORIS (1983) Static response of infrared neurons of crotaline snakes: Normal distribution of interspike intervals. Cell. Mol. Neurobiol. *3*: 27–37.

TERASHIMA, S., R. C. GORIS, and Y. KATSUKI (1968) Generator potential of crotaline snake infrared receptor. J. Neurophysiol. *31*: 682–688.

TERASHIMA, S., R. C. GORIS, and Y. KATSUKI (1970) Structure of warm fiber terminals in the pit membrane of vipers. J. Ultrastruct. Res. *31*: 494–506.

TERASHIMA, S., and P.-J. JIANG (1993) The effect of temperature change on the number of vesicles and on the form of mitochondria in free nerve endings. J. Physiol. Soc. Japan. *55*: 64–65. [In Japanese]

WARREN, J. W., and U. PROSKE (1968) Infrared receptors in the facial pits of the Australian python, *Morelia spilotes*. Science *159*: 439–441.

WELDON, P., M. BACHOO, and C. POLOSA (1990) Depletion by preganlionic stimulation and post-stimulus recovery of large dense core vesicles in synaptic boutons of the cat superior cervical ganglion. Brain Res. *516*: 341–344.

WOOD, R. L. (1959) Intercellular attachment in the epithelium of *Hydra* as revealed by electron microscopy. J. Biophys. Biochem. Cytol. *6*: 343–352.

Calcitonin gene-related peptide immunoreactivity in the trigeminal ganglion of *Trimeresurus flavoviridis*

Miwako Sekitani-Kumagai*[a], Tetsuo Kadota[a], Richard C. Goris[a], Toyokazu Kusunoki[a], Shin-ichi Terashima[b]

[a]*Department of Anatomy, Yokohama City University School of Medicine, Fukuura 3-9, Kanazawa-ku, Yokohama, 236 Japan*
[b]*Department of Physiology, University of the Ryukyus School of Medicine, Nishihara-cho, Okinawa, 903-01 Japan*

Abstract

Crotaline snakes, which have infrared-sensitive pit organs, provide a good model for linking neuron morphology with sensory modality. In the trigeminal ganglion of the habu, *Trimeresurus flavoviridis*, cells positive for calcitonin gene-related peptide-like (CGRP) immunoreactivity were found to be of two types, darkly stained and lightly stained. They were pseudo-unipolar, having an axon divided into stem, peripheral branch, and central branch, all of which were 1 μm or less in diameter. Other, CGRP-negative cells in the ganglion were also pseudo-unipolar, but much larger. In configuration, some of the positive cells were similar to the neurons with A-delta fibers, and others to the neurons with C fibers that have been reported by other workers. On the basis of their distribution and density, and physiological studies by other workers, the CGRP-positive cells were judged to be not part of the infrared-receptive system, but to be involved in the transmission of nociception in small fibers.

Keywords: Calcitonin gene-related peptide-like; Trigeminal ganglion; Crotaline snakes; A-delta fibers; C fibers; Cell size

1. Introduction

Crotaline snakes ('pit vipers') have a pair of pit organs in the loreal region of the face which serve as infrared receptors, aiding in the location and capture of warm-blooded prey. The pit organs, or pits, are innervated mainly by the maxillary branches, and partly by the ophthalmic branch, of the trigeminal nerve; these branches also transmit other modalities of skin sensation (Lynn, 1931; Kishida et al., 1982).

In the brain, infrared sensation is transmitted to the optic tectum via an independent system (Molenaar, 1978a,b; Kishida et al., 1980; Molenaar and Fizaan-Oostveen, 1980; Newman et al., 1980; Meszler et al., 1981).

By studying the neurons in the ganglions of these nerves, it seems possible to determine which cells pertain to which sensory modality. Terashima and Liang (1991, 1994a,b) and Liang and Terashima (1993) used

* Corresponding author.

physiological methods to identify neurons which respond to temperature and mechanical stimulation, temperature stimulation alone, and nociception, and determined that these were all medium-sized pseudounipolar neurons.

In the present work we studied the relationship between the physiological data and neurons containing calcitonin gene-related peptide-like (CGRP) immunoreactivity, because CGRP may be involved in the transmission of various sensory modalities, including pain.

2. Materials and methods

2.1. Immunohistochemistry

For these experiments we used oriental pit vipers (habu, *Trimeresurus flavoviridis*) which had been kept in captivity for some time. Three of these animals weighing 250–450 g were anesthetized with halothane and perfused with heparinized (1 I.U./ml) saline containing 0.05% of an additional anesthetic, tricaine methanesulfonate. This was followed by perfusion with Zam-

boni's fixative (0.2% picric acid and 4% paraformaldehyde in 0.1 M phosphate buffer (PB), pH 7.4). The
right and left trigeminal ganglia were removed and immersed in the same fixative for an additional 3 h at 4°C.
The specimens were then transferred to 30% sucrose in
PB after washing with PB, and kept overnight at 4°C.
Serial horizontal cryostat sections were cut at 40 μm and
mounted on gelatin-coated slides. After gentle drying
with an electric fan, the slides were dipped in the same
fixative solution for 1 h. Then they were washed in
several changes of 0.9% NaCl and 0.3% Triton X-100 in
PB (PBST) and kept at 4°C overnight. The slides were
treated with a protein blocking agent (Immunon) for
suppressed non-specific staining and transferred to
rabbit polyclonal antibodies against CGRP (Cambridge
Research Biochemicals) diluted to 1:1500 with 1% nor-

mal goat serum (NGS), 0.2% bovine serum albumin
(BSA) and 0.1% sodium azide in PBST, and kept in the
solution for 2 days at 4°C. After rinsing in three changes
of PBST for 1 h, the sections were transferred for 3 h to
goat anti-rabbit IgG (Cappel) diluted to 1:200 with the
same solution at room temperature. Next, the slides
were rinsed with three changes of 0.9% NaCl in PB
(PBS), transferred for 90 min to rabbit peroxidase-
antiperoxidase complex (Jackson) diluted to 1:200 with
1% NGS, 0.2% BSA and 0.1% thimerosal in PBS, and
rinsed in three changes of PBS for 1 h. The slides were
transferred to 0.1 M Tris–HCl buffer at pH 7.8 for 10
min and peroxidase reactivity was demonstrated with
the 3,3′-diaminobenzidine method of Graham and Kar-
novsky in the same buffer. The specimens were then
dehydrated in an ethanol series, cleared with xylene, and
mounted with Canada balsam. Preabsorption of the
antibody with 50 μM rat α-CGRP (Rat) (Peptide Institute) abolished immunostaining.

2.2. Morphometry

Sketches of the cell bodies in four sections at a
magnification of about 248 times were made with a
camera lucida. These sketches were used with an image
processor (Nippon Avionics, IC5098) to measure the

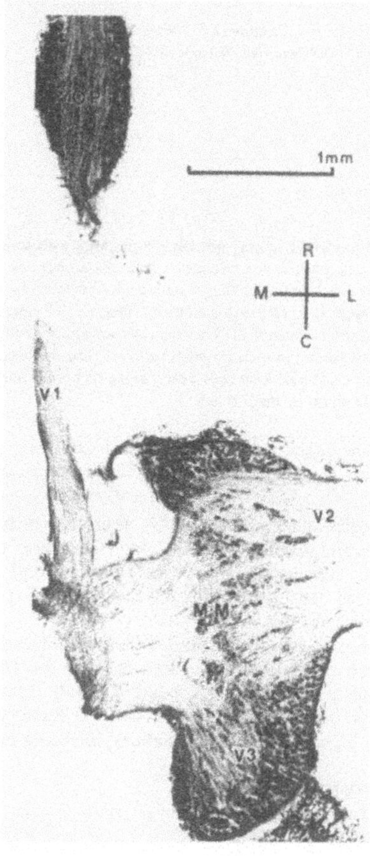

Fig. 1. Horizontal section of the trigeminal ganglion, right side, CGRP
immunoreactivity. OP, ophthalmic ganglion; MM, maxillo-
mandibular ganglion; V1, ophthalmic nerve; V2, maxillary nerve; V3,
mandibular nerve.

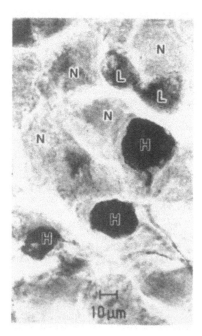

Fig. 2. Cells in the maxillo-mandibular ganglion showing positive or
negative CGRP immunoreactivity. H, darkly stained cells; L, lightly
stained cells; N, negative cells.

sectional area, the long and short diameter of the neurons considered to be ellipsoids, and the diameter of the same neurons considered to be spheres, in order to be able to make comparisons with the data of other workers, who often use differing methods of measurement. The data were analyzed with a non-parametric test.

Spatial distribution of CGRP-positive cells within the ganglion was observed with an image processor (TRI, Ratock System engineering, Japan). Enlarged figures, representing outlines of the ganglion with dots showing positive perikarya, were drawn with a camera lucida from 40-μm horizontal serial sections. From these figures the image processor made two files along the z

axis, one upward from the most expanded part of the tissue and the other downward. A 3-dimensional image was obtained by combining the two files into a third file. The system was then used to estimate the cell density in space.

3. Results

As shown in Fig. 1, CGRP-positive and CGRP-negative cells were seen in both the ophthalmic and maxillo-mandibular ganglia. We treated the maxillo-mandibular ganglion as a single entity, since we could not distinguish the maxillary division from the mandibular division.

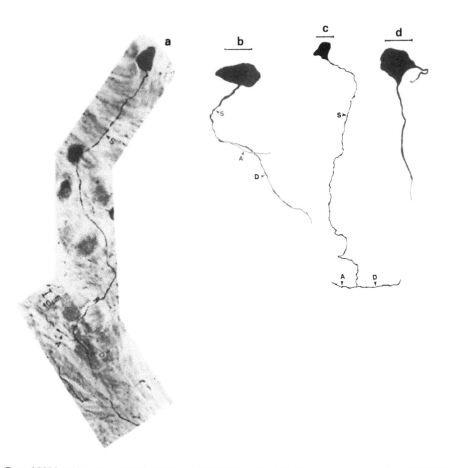

Fig. 3. Types of CGRP-positive cells. A, central branch; D, peripheral branch; S, stem. (a) A-delta fiber type neuron, where A was thinner than S and D, and branched like a T. (b) A-delta type neuron with stem branched like a Y. (c) C-fiber type neuron with thin processes of equal thickness. (d) A rare bipolar cell. Bar in b, c, and d indicates 20 μm.

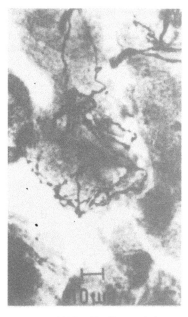

Fig. 4. Arborization of CGRP-positive fibers enveloping a negative cell.

3.1. CGRP-positive cells

CGRP-positive cell bodies and their processes showed homogeneous dark brown staining. These were nearly round, ovoid cells, of two types: darkly stained and

lightly stained (Figs. 1 and 2). It was impossible to tell if these were split cells, or actually different cells, so we measured them as split cells. As shown in Fig. 3, most of the positive cells were pseudo-unipolar cells, with an axon having a stem, a central branch, and a peripheral branch. Most of the stems issued straight out from the cell body, although there were also a few with a crooked shape. The point where the stem divided into the central branch and the peripheral branch was either T-shaped or Y-shaped, and was swollen in some neurons. These three processes were all less than 1 μm in diameter, the stem being 0.5–1 μm, the central branch less than 0.5 μm, and the peripheral branch 0.5–1 μm, with some peripheral branches less than 0.5 μm.

These cells could be divided into neurons where the central branch was thinner than the stem and peripheral branch (similar to the slow-conducting A-delta neurons of Liang and Terashima (1993)) (Fig. 3a, b), and into neurons where all three processes were thin (similar to the C fibers of Terashima and Liang (1994b)) (Fig. 3c). Occasionally, atypical bipolar cells were also seen (Fig. 3d). There were also a few CGRP-positive fiber arborizations with varicosities enveloping CGRP-negative cell bodies (Fig. 4), but it was impossible to trace the origin of these fibers.

In horizontal sections of the central and peripheral neural roots originating from the trigeminal ganglion, the ratio of CGRP-immunoreactive nerve fibers with a diameter of less than 0.5 μm/104 μm^2 was 78% for the central root and 31% for the peripheral root of the ophthalmic ganglion (six preparations); and 29% for the maxillary division (six preparations), 51% for the mandibular division (three preparations) and 96.4% for the

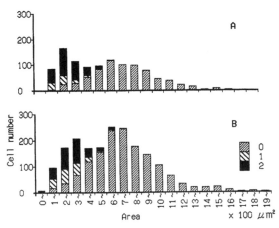

Fig. 5. Histogram of the area of cells. 0, negative; 1, lightly stained; 2, darkly stained. (A) Ophthalmic ganglion showing two peaks. (B) Maxillo-mandibular ganglion showing two peaks.

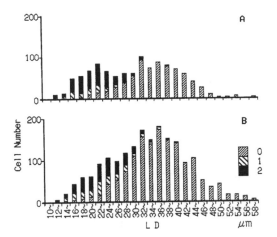

Fig. 6. Histogram of the long diameter of cells. 0, negative; 1, lightly stained; 2, darkly stained. (a) Ophthalmic ganglion showing two peaks. (b) Maxillo-mandibular ganglion.

central nerve root (six preparations, Fig. 7). Nerve fibers tended to become thinner as the distance from the ganglia increased.

3.2. Measurement of all sections of cell bodies

A significant difference was apparent in cell size between CGRP-positive and CGRP-negative cells (Figs. 2 and 5). There were two peaks in the histogram of the cell area in the ophthalmic ganglion, one at 200–300 μm^2 and the other at 600–700 μm^2 (Fig. 5A), and in the maxillo-mandibular ganglion, at 300–400 μm^2 and 700–800 μm^2 (Fig. 5B). The smaller size groups were composed mainly of CGRP-positive cells, and the larger groups of CGRP-negative cells. A similar pattern was observed in the histogram of the cell long diameter in the ophthalmic ganglion, but it was not distinct in the maxillo-mandibular ganglion (Fig. 6). Table 1 shows the mean data for the cell area, long diameter, short diameter, ratio of long to short diameter, and putative circular diameter for each group of cells. The size of CGRP-negative cells in the ophthalmic ganglion, including split cells (Table 1a), was 36.6×26.2 μm, and their area was 773.02 μm^2. In the maxillo-mandibular ganglion, cell size was 37.8×25.6 μm, and area was 782.35 μm^2; these were significantly greater in long/short diameter than cells in the ophthalmic ganglion, so that they appeared larger and more elongate. In these cells darkly and lightly stained types could not be clearly identified, so they were all treated as belonging to the same group.

In CGRP-positive cells it was difficult to identify the nucleolus because reaction products covered the cell nu-cleus. If we suppose that lightly and darkly stained CGRP cells belong to the same group (Table 1c), then the figures are 22.2×15.6 μm and 280.08 μm^2 for the ophthalmic ganglion and 24.2×16.0 μm and 313.80 μm^2 for the maxillo-mandibular ganglion.

3.3. Measurement of CGRP-negative cells containing a nucleolus

CGRP-negative cells containing a nucleolus were considered close to the true form of the cell, and we excluded split cells which did not contain a nucleolus (Table 1b). In the ophthalmic ganglion, the area of cells containing a nucleolus (42.3×31.9 μm) was 1081.89 μm^2; the same figures for the maxillo-mandibular ganglion were 44.9×30.6 μm, and 1093.24 μm^2, respectively. There was no significant difference in cell area, but the cells in the maxillo-mandibular ganglion seemed to be slightly more elongate (Table 1b).

3.4. Abundance of CGRP cells

In four sections each of the ophthalmic and maxillo-mandibular ganglia, 33.3% and 23.3%, respectively, of the cells were CGRP-positive. This showed a significant difference using χ^2 test (Table 2).

3.5. Distribution in space

In the ophthalmic ganglion, both CGRP-positive and CGRP-negative cells tended to aggregate at the surface of the ganglion, though no specific distribution could be observed (Fig. 1). CGRP-positive cell density was $112.6 \times 10^3/mm^3$ at the boundary of the ganglion and $46.9 \times 10^3/mm^3$ at the center (Table 3a). Since the

Table 1
Measurements of cell bodies of neurons

Table 1a
CGRP negative cells[a]

	V1 ($n = 720$)	V2+3 ($n = 1494$)
Area	773.02 ± 281.33	782.35 ± 307.88
LD	36.6 ± 7.0	37.8 ± 8.2*
S.D	26.2 ± 6.0	25.6 ± 5.9*
L/D	1.4 ± 0.3	1.5 ± 0.3*
D	30.8 ± 5.8	31.0 ± 6.2

Table 1b
CGRP negative cells with nucleolus

	V1 ($n = 238$)	V2+3 ($n = 113$)
Area	1081.89 ± 354.98	1093.24 ± 332.86
LD	42.3 ± 8.2	44.9 ± 8.1*
S.D.	31.9 ± 5.5	30.6 ± 5.1*
D	36.6 ± 6.1	36.9 ± 5.4

Table 1c
Lightly and heavily stained CGRP positive cells[a]

	V1 ($n = 360$)	V2+3 ($n = 454$)
Area	280.08 ± 106.31	313.8 ± 132.6*
LD	22.2 ± 4.7	24.2 ± 6.0*
S.D.	15.6 ± 3.4	16.0 ± 3.6
D	18.5 ± 3.5	19.6 ± 4.1*

Values show the mean ± S.D. determined according to Mann-Whitney statistical test. Area, cross sectional area of the cells (μm^2); D, diameter of cells treated as spheres (μm); LD and SD, long and short diameters of cells treated as ellipsoids (μm); n, number of cell sections; L/D, long diameter/short diameter; V1, ophthalmic ganglion; V2+3, maxillo-mandibular ganglion.
[a]including split cells (see text).
*Significant difference ($P \leq 0.05$).

center of the ganglion contains many fibers, the cell density was lower there than at the circumference.

Ganglion cells could be divided into six groups, according to whether they were inside or outside fiber bundles, i.e., the rostral and central part of the origin of the nervus maxillaris and the lateral and medial part of the origin of the nervus mandibularis (Fig. 1), and the dorsal and ventral margin of the ganglion (Table 3). Cell density was low in the central part which contained

Table 3
Positive cell density in space

Table 3a
Ophthalmic ganglion

Boundary	112. 6 × 10³/mm³
Center	46.9 × 10³/mm³
(Mean	79.8 × 10³/mm³)

Table 3b
Maxillo-mandibular ganglion

Rostral part of the origin of maxillary n.	48.3 × 10³/mm³
Central part of the origin of maxillary n.	34.9 × 10³/mm³
Lateral part of the origin of mandibular n.	58.6 × 10³/mm³
Internal part of the origin of mandibular n.	64.1 × 10³/mm³
Dorsal margin of the ganglion	44.3 × 10³/mm³
Ventral margin of the ganglion	44.1 × 10³/mm³
(Mean	49.1 × 10³/mm³)

many nerve bundles, as shown in Table 3b and Fig. 1. Three-dimensional density varied from 35×10^3 to $64 \times 10^3/mm^3$ (Table 3b).

4. Discussion

4.1. Distribution and density of CGRP-immunoreactive cells

In four horizontal sections, 33.3% of the cells in the ophthalmic ganglion and 23.3% in the maxillo-mandibular ganglion showed CGRP immunoreactivity. The difference is statistically significant. The density/volume of the ganglion was also higher in the ophthalmic ganglion ($79.8 \times 10^3/mm^3$ on average) than in the maxillo-mandibular ganglion ($49.1 \times 10^3/mm^3$ on average). We presume that the number of CGRP-immunoreactive cells is not directly concerned with the innervation of the pit organ, because such cells were fewer in the maxillo-mandibular ganglion, which innervates the major part of the pit organ. CGRP-immunoreactive neurons were also observed in the mandibular division of this ganglion, which does not innervate the pit organ.

Table 2
Percentage of CGRP negative and positive cells

	No. of negative cells	No. of positive		Negative cells (%)	Positive cells (%)
		cells	sum		
V1	720	360	1080	66.7	33.3
V2+3	1494	454	1948	76.7	23.3

The percentage of the two types was significantly different (χ^2-test: $P < 0.01$)

The coexistence of CGRP and substance P-like immunoreactivity in the same neurons has been confirmed in the mammalian dorsal root and trigeminal ganglia, where 40–50% of all cells have substance P-like immunoreactivity (Lee et al., 1985, 1986; Ju et al., 1987; O'Brien et al., 1989; Carr et al., 1990). In snakes, about 30% of all ganglionic cells are small and have substance P-like immunoreactivity in *Agkistrodon blomhoffii* (Kadota et al., 1988). A similar abundance and size of CGRP cells has also been shown in *T. flavoviridis*. Therefore, it appears that the coexistence of CGRP and substance P-like immunoreactivity is a common characteristic of pit vipers in general.

According to Terashima (1987), fibers containing substance P-like immunoreactivity do not serve only in the transmission of nociceptive impulses, but may also act as modulators of the infrared sensory system. If so, CGRP-positive cells may also be involved in pain transmission and modulation of the infrared system.

4.2. Size and modality of trigeminal ganglion neurons

Terashima and Liang (1991) observed the morphology of trigeminal neurons in *T. flavoviridis* by intracellular injection of horseradish peroxidase (HRP), after physiologically determining the thermal or mechanical sensitivity of the pseudounipolar neurons. Temperature-sensitive cells were 42.1 ± 5.3 μm in major diameter and 31.2 ± 4.7 μm in minor diameter, $n = 9$, whereas cells with both thermal and mechanical sensitivity were smaller, being 40.6 ± 3.1 μm in major diameter and 26.0 ± 2.8 μm in minor diameter, $n = 5$. Nociceptive A-delta neurons of the fast conducting type (11.2 ± 2.0 m/s) were 47.8 ± 6.2 μm in major diameter and 36.9 ± 5.4 μm in minor diameter, and those of the slow conducting type (3.8 ± 1.1 m/s) were 44.3 ± 6.3 μm in major diameter and 33.8 ± 4.2 μm in minor diameter (Liang and Terashima, 1993). These sensory cells probably pertain to the CGRP-negative group, because the size of CGRP-negative cells containing a nucleolus in our preparations of the maxillo-mandibular ganglion was 44.9 ± 4.1 μm in long diameter and 30.6 ± 5.1 μm in short diameter, values similar to those obtained by Terashima and Liang (1991) and Liang and Terashima (1993).

Terashima and Liang (1994b) studied two C mechanical nociceptive cells intracellularly labeled with HRP in the maxillo-mandibular ganglion, and found that they measured 22×20 μm and 20×18 μm, which is within the mean ± S.D. cell size of CGRP-positive cells in our maxillo-mandibular preparations. Thus, on the basis of the data in the histograms of Figs. 5 and 6, we surmised that C mechanical nociceptive cells correspond to our small, CGRP-positive cells, and thermoreceptive and mechanoreceptive cells to our large, CGRP-negative cells.

In mammals, in the dorsal root ganglion of the rat, CGRP is found in 46% of C-fiber neurons (< 1.3 m/s), 33% of A-delta fiber neurons ($2–12$ m/s), and 17% of A-alpha/delta neurons (> 12 m/s) (McCarthy and Lawson, 1990).

Kishida et al. (1982) used retrograde HRP labeling from the pit organ to the trigeminal ganglion, and found that labeled ganglion cells measured $20–39$ μm in mean diameter. In our present work, CGRP-negative cells were 31.0 ± 6.2 μm in diameter when treated as spheres, and 36.9 ± 5.4 μm when the cells measured contained a nucleolus. Thus, the cells measured by Kishida et al. (1982) included the CGRP-negative group.

4.3. Branching patterns of the cell processes

The diameter of CGRP-immunoreactive fibers was either $0.5–1$ μm or less than 0.5 μm, as stated above. Terashima and Liang (1991, 1994a,b) were of the opinion that the branching pattern of the three processes was a key indicator of the sensory modality of the neurons. Cells with A-delta fibers as described by Liang and Terashima (1993) were larger than CGRP-immunoreactive cells, although CGRP-positive cells had the same morphological pattern, i.e., the central branch was thinner than (A-delta fibers), or of equal diameter to (C fibers), the stem and peripheral branch. Our results coincided with theirs regarding the thickness of the three processes and the fact that the diameter of the central branch was equal to (C fibers), or thinner than (A-delta fibers), that of the other two processes.

It is reasonable to think that CGRP-immunoreactive cells are involved in the transmission of nociception, but not in the transmission of infrared stimulation, because no CGRP-immunoreactive fibers could be observed in the pit organs themselves or in the nucleus of the lateral descending trigeminal tract (*A. blomhoffi*, unpublished observation). As shown in Fig. 7, the central root of the trigeminal ganglion contains mainly thin CGRP fibers, whereas the peripheral root contains relatively many thick CGRP fibers. Since neurons with thick peripheral branches in the peripheral root of the trigeminal ganglion and thin central branches in its central root have been identified as slow-conducting A-delta nociceptor cells (Liang and Terashima, 1993), it is highly probable that the CGRP-immunoreactive cells which we observed, although of smaller size than those of Liang and Terashima (1993), are slow-conducting A-delta mechanical nociceptor cells because they show the same axonal morphology.

The observed combination of thin peripheral branches in the peripheral ganglion root and thin central branches in its central root can be deduced as belonging to C fiber neurons. These cells also include mechanical nociceptors (Terashima and Liang, 1993). There appears to be a strong relationship between CGRP and mechanical nociceptors. For example, Hoheisel et al. (1994)

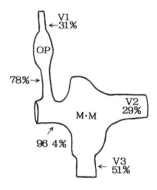

Fig. 7. Percentage of thin fibers among CGRP-positive fibers within a square of 10.2 μm (104 μm²). OP, ophthalmic ganglion; MM, maxillo-mandibular ganglion; V1, ophthalmic nerve; V2, maxillary nerve; V3, mandibular nerve. Horizontal section.

reported the possibility that CGRP-positive cells in the lumbar dorsal root of the rat were nociceptors, and Beckers et al. (1992) theorized that CGRP cells are involved in ocular protecting responses to noxious stimuli.

4.4. Pericellular arborization

According to Scharf (1958), the pericellular arborization by fibers in spinal ganglia was first found by Ehrlich (1886) and Aronson (1886) using methylene blue staining. Immunohistochemically, pericellular arborization with substance P-like immunoreactivity is found in the jugular and nodose ganglia of rabbits and pigeons (Katz and Karten, 1980). Later, Kuwayama and Stone (1986) found a similar structure composed of fibers containing CGRP, substance P, and cholecystokinin in the trigeminal ganglion of the guinea pig. Although rare, we found similar structures in our Trimeresurus, where peptide-positive arborization with varicosities enveloped a peptide-free cell body. This is the first report of such a structure in a reptile.

4.5. Nerve cell bodies in the ophthalmic and maxillo-mandibular ganglia

In our material, there was a significant difference in CGRP-immunoreactive cell size between the ophthalmic and maxillo-mandibular ganglia, i.e., 280.08 ± 106.31 μm² vs. 313.8 ± 132.6 μm², respectively. On the other hand, no significant difference was found between CGRP-negative cells in the two ganglia. The reason for this is not clear.

CGRP-positive cells in the trigeminal ganglion of rats can be divided into a large (40 μm in average diameter) and small (20 μm in average diameter) (Lee et al., 1985). In the work by Hoheisel et al. (1994), CGRP-positive

cells ranged from 201 to 1179.2 μm². The average diameter of cells treated as spheres in our material was 19 μm, but there were no 'large' cells corresponding to those of the rat.

4.6. Conclusions

To conclude, whether the size difference described here is the result of an evolutionary process must await further research. However, it can be said with a strong degree of probability that CGRP appears to be a transmitter for nociception in small fibers.

Acknowledgement

We wish to thank Dr. Shosaku Hattori of the Amami Branch Laboratory, Institute of Medical Science, University of Tokyo, for providing the snakes used in this work.

References

Aronson, H. (1886) Beiträg zur Kenntnis der zentralen und peripheren Nervenendigungen. Inaug. Diss. Berlin, (cited from Scharf, 1958).

Beckers, H.J.M., Klooster, J., Vrensen, G.F.J.M. and Lamers, W.P.M.A. (1992) Ultrastructural identification of trigeminal nerve endings in the rat cornea and iris. Invest. Ophthalmol. Vis. Sci., 33: 1979–1986.

Carr, P.A., Yamamoto, T. and Nagy, J.I. (1990) Calcitonin gene-related peptide in primary afferent neurons of rat: Co-existence with fluoride-resistant acid phosphatase and depletion by neonatal capsaicin. Neuroscience, 36: 751–760.

Ehrlich, P. (1886) Über die Methylenblaureaction der lebenden Nervensubstanz. Dtsch. Med. Wschr., 49–52, (cited from Scharf, 1958).

Hoheisel, U., Mense, S. and Scherotzke, R. (1994) Calcitonin gene-related peptide-immunoreactivity in functionally identified primary afferent neurones in the rat. Anat. Embryol., 189: 41–49.

Ju, G., Hökfelt, T., Brodin, E., Fahrenkrug, J., Fischer, J.A. and Frey, P. (1987) Primary sensory neurons of the rat showing calcitonin gene-related peptide immunoreactivity and their relation to substance P-, somatostatin-, galanin-, vasoactive intestinal polypeptide- and cholecystokinin-immunoreactive ganglion cells. Cell Tissue Res., 247: 417–431.

Kadota, T., Kishida, R., Goris, R.C. and Kusunoki, T. (1988) Substance P-like immunoreactivity in the trigeminal sensory nuclei of the an infrared-sensitive snake, Agkistrodon blomhoffi. Cell Tissue Res., 253: 311–317.

Katz, D.H. and Karten, H.J. (1980) Substance P in the vagal sensory ganglia: Localization in the cell bodies and pericellular arborizations. J. Comp. Neurol., 193: 549–564.

Kishida, R., Amemiya, F., Kusunoki, T. and Terashima, S. (1980) A new tectal afferent nucleus of the infrared sensory system in the medulla oblongata of crotaline snakes. Brain Res., 195: 271–279.

Kishida, R., Terashima, S., Goris, R.C. and Kusunoki, T. (1982) Infrared sensory neurons in the trigeminal ganglia of crotaline snakes: Transganglionic HRP transport. Brain Res., 241: 3–10.

Kuwayama, T. and Stone, R.A. (1986) Neuropeptide immunoreactivity of pericellular baskets in the guinea pig trigeminal ganglion. Neurosci. Lett., 64: 169–172.

Lee, Y., Kawai, Y., Shiosaka, S., Takami, K., Kiyama, H., Hillyard, C.J., Girgis, S., MacIntyre, I., Emson, P.C. and Tohyama, M. (1985) Coexistence of calcitonin gene-related peptide and substance P-like peptide in single cells of the trigeminal ganglion of the rat: immunohistochemical analysis. Brain Res., 330: 194–196.

Lee, K.H., Chung, K., Chung, J.M. and Coggeshall, R.E. (1986) Correlation of cell body size, and signal conduction velocity for individually labelled dorsal root ganglion cells in the cat. J. Comp. Neurol., 243: 335–346.

Liang, Y.F. and Terashima, S. (1993) Physiological properties and morphological characteristics of cutaneous and mucosal mechanical nociceptive neurons with A-delta peripheral axons in the trigeminal ganglia of crotaline snakes. J. Comp. Neurol., 328: 88–102.

Lynn, W.G. (1931) The structure and function of the facial pit of the pit vipers. Am. J. Anat., 49: 97–139.

McCarthy, P.W. and Lawson, S.N. (1990) Cell type and conduction velocity of rat primary sensory neurons with calcitonin generelated peptide-like immunoreactivity. Neuroscience, 34: 623–632.

Meszler, R.M., Auker, C.R. and Carpenter, D.O. (1981) Fine structure and organization of the infrared receptor relay, the lateral descending nucleus of the trigeminal nerve in pit vipers. J. Comp. Neurol., 196: 571–584.

Molenaar, G.J. (1978a) The sensory trigeminal system of a snake in the possession of infrared receptors I. The sensory trigeminal nuclei. J. Comp. Neurol., 179: 123–136.

Molenaar, G.J. (1978b) The sensory trigeminal system of a snake in the possession of infrared receptors II. The central projections of the trigeminal nerve. J. Comp. Neurol., 179: 137–152.

Molenaar, G.J. and Fizaan-Oostveen, J.L.F.P. (1980) Ascending projection from the lateral descending and common sensory trigeminal nuclei in python. J. Comp. Neurol., 189: 555–572.

Newman, E.A., Gruberg, E.R. and Hartline, P.H. (1980) The infrared trigemino-tectal pathway in the rattlesnake and in the python. J. Comp. Neurol., 191: 465–477.

O'Brien, C., Woolf, C.J., Fitzgerald, M., Lindsay, R.M. and Molander, C. (1989) Difference in the chemical expression of primary afferent neurons which innervate skin, muscle or joint. Neuroscience, 32: 493–502.

Scharf, J.H. (1958) Zur Cytologie der sensiblen Ganglien. In: W. v. Möllendorff und W. Bargmann (Eds.), Handbuch der mikroskopischen Anatomie des Menschen, IV Nervensystem; 3 sensiblen Ganglien, Springer, Berlin, pp. 341–349.

Terashima, S. (1987) Substance P-like immunoreactive fibers in the trigeminal sensory nuclei of the pit viper, *Trimeresurus flavoviridis*. Neuroscience, 23: 685–691.

Terashima, S. and Liang, Y-F. (1991) Temperature neurons in the crotaline trigeminal ganglia. J. Neurophysiology, 66: 623–634.

Terashima, S. and Liang, Y-F. (1993) Modality difference in the physiological properties and morphological characteristics of the trigeminal sensory neurons. Jpn. J. Physiol., 43: Suppl. 1, 267–274.

Terashima, S. and Liang, Y-F. (1994a) Touch and vibrotactile neurons in a crotaline snake's trigeminal ganglia. Somatosens. Mot. Res., 11: 169–181.

Terashima, S. and Liang, Y-F. (1994b) C mechanical nociceptive neurons in the crotaline trigeminal ganglia. Neurosci. Lett., 179: 33–36.

Distinct Morphological Characteristics of Touch, Temperature, and Mechanical Nociceptive Neurons in the Crotaline Trigeminal Ganglia

YUN-FEI LIANG, SHIN-ICHI TERASHIMA, AND AI-QING ZHU
Department of Physiology, University of the Ryukyus School of Medicine,
Okinawa 903-01, Japan

ABSTRACT

Intrasomal recording and horseradish peroxidase injection techniques were employed in vivo to determine the morphological characteristics of touch, temperature, and mechanical nociceptive neurons in the trigeminal ganglia of crotaline snakes. The touch neurons, with a peripheral axon conducting at the A–β range, could be subdivided into tactile and vibrotactile neurons according to their response properties, but there were no morphological differences between them. These neurons exhibited a large and oval soma and possessed a set of large stem, peripheral, and central axons which were all myelinated and equal in diameter with a constriction at the bifurcation. The temperature neurons, which conducted peripherally at the A–δ range, were physiologically separated into thermosensitive and thermo-mechanosensitive neurons, which were also morphologically indistinguishable. The temperature neurons had a round soma of medium size and a set of medium axons with varied axonal bifurcation patterns. All axons of these neurons were myelinated, but the central axon was thinner than the stem and peripheral axons. The mechanical nociceptive neurons, which had a peripheral axon conducting at the A–δ range, were morphologically heterogeneous based on their conduction velocities. The neurons conducting at the fast A–δ range were morphologically similar to the temperature neurons in the ganglion excepting their thinner central axons, whereas those at the slow A–δ range had a thinner myelinated stem axon that gave rise to a thinner myelinated peripheral axon and an unmyelinated central axon with a bifurcation of either a triangular expansion at the bifurcating point or a central axon arising straightforwardly from the constant stem and peripheral axons. This study revealed that distinct morphological characteristics do exist for the touch and temperature neurons and the subtypes of mechanical nociceptive neurons in the trigeminal ganglion, but not for the subfunctional types of touch neurons or temperature neurons. © 1995 Wiley-Liss, Inc.

Indexing terms: functional morphology, intracellular horseradish peroxidase, soma area, axon diameter, axonal bifurcation

The variety in soma size, axon diameter, and axonal bifurcation pattern of the primary sensory neurons in the ganglia of human and various vertebrate animals has been well documented (Dogiel, 1898; Hatai, 1902; Cajal, 1906; Ranson, 1912; Ranson and Davenport, 1931; Gasser, 1955; Ha, 1970; Lieberman, 1976). Combining intracellular recording with intracellular injection of a marker, such morphological variety of the sensory ganglion neurons has recently been directly related to peripheral nerve conduction velocity in individual neurons (Yoshida and Matsuda, 1979; Harper and Lawson, 1985; Cameron et al., 1986; Lee et al., 1986; Hoheisel and Mense, 1987; Mense, 1990; Nagy et al., 1993). Hoheisel and Mense (1987) revealed a particu-

larly detailed relationship between morphological characteristics and conduction velocity for the neurons in the dorsal root ganglion, in which it was reported that the neurons conducting in the A–β range had thick axons, whereas axons of neurons in the A–δ range were different according to their conduction velocities: i.e., neurons conducting in the fastest range had myelinated axons on both sides of the bifurcation; those conducting in the intermediate range had a myelinated peripheral axon and an unmyelinated central axon; and those conducting in the lower range had unmyelinated axons on both peripheral and central sides. These results aroused interest as to whether such morphological characteristics relate to sensory modalities.

Although conduction velocity could usually be utilized to suggest a probable functional cell type (Burgess and Perl, 1973), such suggestions have obvious shortcomings because some neurons being of different response modalities could conduct at the same velocity and some others being of the same response modality may fall into different ranges of conduction velocity (see the results below). Therefore, even though the conduction velocities of the ganglion neurons have been known to relate to their morphologies to some degree (Hoheisel and Mense, 1987; Mense, 1990), neuronal morphologies cannot be reliably connected with their sensory modalities by conduction velocities. Thus direct approaches concerning the relationship of morphological characteristics of the ganglion neurons to their sensory modality are of particular significance, because knowing of such structure-functional correlation could be helpful in understanding the peripheral mechanisms of sensory processing. Some such attempts have been made on the dorsal root ganglion (Cameron et al., 1986) and trigeminal ganglion (Jacquin et al., 1986, 1992). However, in the experiment of Cameron and co-workers (1986), only soma size was studied with intracellular injection of Lucifer yellow, and in those of Jacquin et al. (1986, 1992), the number of the neurons labeled with horseradish peroxidase (HRP) or Neurobiotin was too small to evaluate the relation of neuronal morphology with sensory modality in the ganglion. As it is difficult to keep a sufficient stable intracellular recording in vivo for defining the cell response modality, the variety in soma size, axon diameter, and axonal bifurcation pattern of primary ganglion neurons has not been systematically correlated with the sensory modality.

Concerning the relationships of soma size to peripheral axon diameter and conduction velocity, studies with intracellular injection of marker in individual neurons have yielded controversial results. For example, Gallego and Eyzaguirre (1978) did not observe significant size difference between A and C neurons in the nodose ganglion, whereas other researchers reported a linear relation between soma size and peripheral axon diameter and/or conduction velocity in the dorsal root ganglion (Yoshida and Matsuda, 1979; Cameron et al., 1986). Lee and his co-workers (1986) found a significant relationship for A neurons but not for C neurons in the dorsal root ganglion. In contrast, Harper and Lawson (1985) demonstrated such a relationship for C neurons and also for A–δ neurons. In the dorsal root ganglion as well, Hoheisel and Mense (1987) reported a linear relation between soma size and the peripheral conduction velocity when the neurons of all groups were considered together, but not for A–δ neuron or C neuron group alone. However, little is known of such relationships in trigeminal ganglion neurons.

The trigeminal ganglion of crotaline snakes is located laterally to the brainstem not covered by the forebrain and is well protected on all sides by bony projections of the skull. This peculiar topographic location has made it possible to expose the ganglion without injury to the nervous system and to obtain a stable intrasomal recording in vivo (Liang and Terashima, 1993). The present study was designed to examine whether distinct morphological characteristics including soma area, axon diameter, and axonal bifurcation pattern exist for identified touch, temperature, and mechanical nociceptive neurons in the trigeminal ganglion and to assess the relationships of conduction velocity to peripheral axon diameter and soma area and the relation of peripheral axon diameter to the soma area for the individual functional cell types and the overall labeled neurons. Some of the data have been presented in a brief report of a proceeding (Terashima and Liang, 1993).

MATERIALS AND METHODS
Animal preparation

Experiments were performed on 60 crotaline snakes of either sex weighing 250–450 g. Anesthesia was initiated with halothane, and in some experiments where only non-noxious stimuli were used the anesthesia was maintained with ketamine (25 mg/kg, im). The animal was then immobilized with pancuronium (2 mg/kg, im). Artificial respiration of unidirectional moisturized airflow of 0.5 liters/min was applied and some hypodermic needles were inserted into the posterior air sack of the lung to serve as an outlet for the air. Two pairs of our own designed snake head holders were used to hold the animal in place. The trigeminal ganglion was exposed by drilling a hole in the lateral skull through the inner ear on each side. Bilateral mandibular branches and superficial maxillary branches of the trigeminal nerves were exposed at the point about 10 mm and 16 mm from the trigeminal ganglion, respectively, for placement of a pair of bipolar stimulating electrodes which were used to activate the impaled neurons and measure their response latency. A pool was formed on each side by skin flaps and filled with liquid paraffin to moisten the exposed nerves. The heart rate was continuously monitored at 50–90 strokes/min and the microcirculation in the iris was monitored under a dissecting microscope. The pupillary reflex was frequently checked and the room temperature was maintained at 24–26°C. When necessary, supplementary doses of pancuronium were administered during the experiment.

Neuron classification

Trigeminal ganglion neurons were impaled with glass microelectrodes which contained a 4–6% solution of HRP (Sigma type VI) in 0.5 M KCl and 0.05 M Tris-buffer (pH 7.3) and had a DC resistance of 60–100 MΩ. The microelectrodes were connected to the preamplifier of a conventional recording system by Ag-AgCl wire.

Since the temperature neurons showed spontaneous discharges, an increase in size of action potentials with a sudden DC shift indicated a preliminary penetration of a temperature neuron. An ice cube and a heat stimulus of radiation of a human hand was used for further identification of temperature neurons, and an He-Ne laser beam (632.8 nm, 2 mW) focused in diameter of 0.2 mm was used for detecting the receptive field (for details, see Terashima and Goris, 1979; Terashima and Liang, 1991). A set of von Frey hairs was applied to the receptive field that had been located by a laser beam to detect the mechanical sensation. The neurons responsive only to the stimulation of a laser beam were classified as thermosensitive neurons, and those responsive to both laser beam and von Frey hairs with a threshold of less than 10 mg were classified as thermo-mechanosensitive neurons. For the other neurons which had no spontaneous discharges, a sudden large DC shift was taken as the first indication of an intracellular recording. After penetration of a cell an intense square-wave pulse with duration of 100–1,000 μs was applied to the mandibular or the maxillary division of the trigeminal nerves to activate the penetrated cell and then the intensity of electrical stimulation was adjusted to a level just above the threshold for measuring the response latency. When a somal action potential was elicited, the peripheral receptive field was explored with a systematic progression of hand-held stimuli, i.e., touching with a set of von Frey hairs, pressing with a blunt glass probe, vibrating with a specially designed vibrator, warming with a 30–40°C forceps, cooling with an ice cube, and pricking with a syringe needle. The neurons which respond to a von Frey hair of less than 10 mg but failed to follow any stimulating frequency produced by a vibrator were classified as tactile neurons, and those which had a response threshold of less than 10 mg and could follow the stimulating frequency of 2–300 Hz were classified as vibrotactile neurons. The neurons which responded exclusively to a needle pricking of the receptive field were classified as mechanical nociceptive neurons which could be morphologically divided into two subtypes (see Results).

Histological processing

Iontophoretical injection of HRP into the somata was conducted through the recording microelectrodes for 15–20 minutes by using a positive current of 10–30 nA with 1,500 ms duration and 500 ms off-time after physiological identification of neurons. To avoid confusion, only one HRP injection was made into each ganglion. After a survival time of about 24 hours, the animals were perfused through the heart with 200 ml of saline at 10°C followed by 200 ml of fixative containing 2% paraformaldehyde and 2.5% glutaraldehyde in 0.1 M phosphate-buffer (pH 7.4, 10°C) and 200 ml of 10% phosphate-buffer sucrose solution (pH 7.4, 10°C). The trigeminal ganglia were removed and stored in 30% sucrose solution in phosphate-buffer overnight at 4°C. Serial frozen sections were cut at 80 μm in the horizontal plane and processed with a nickel-cobalt diaminobenzidine reaction (Adams, 1981). The sections were mounted on chrome alum-gelatin N coated slides and counterstained with 1% cresyl violet after being air dried.

Data analysis

Using a 40× objective and a camera lucida drawing tube (Olympus BH2-DA), somata of the labeled neurons were drawn under the light microscope (Olympus BH-2) when they were focused in the plane where they had the greatest cross-sectional area. Soma areas were then measured on the drawings, and the data were digitized and analyzed by a microcomputer (NEC PC-9801 VX) with the program Cosmosone 98 (NIPPON KOGAKU KK). The diameters of the labeled stem, peripheral, and central axons were directly measured under the light microscope at 50-μm intervals from the T-bifurcations using 40× and 100× (oil) objectives and an ocular micrometer calibrated with a stage micrometer. Ten measurements were made on peripheral axons and five on central and stem axons, and the mean values were taken as their size. Some of the labeled neurons were reconstructed by means of camera lucida drawings.

The data of measured soma areas and axon diameters are presented as the mean ± SD. One-way analysis of variance was used to compare the effects of different cell types on the soma area, and two-way analysis of variance with unbalanced data was used to test for any interaction effects

between the two factors of cell-type and axon-type. Since all these tests revealed statistical significances, they were followed by the Student-Newman-Keuls test for all pairwise comparisons. Linear regression and correlation analyses were applied to examine the relations between soma area and conduction velocity, axon diameter and conduction velocity, and soma area and axon diameter for the overall labeled neurons and the individual functional groups. For all statistical evaluations, P values < .05 were considered to be significant.

RESULTS
Response characteristics

Ninety-one trigeminal ganglion neurons innervating the oral mucosa or facial skin with a peripheral axon conducting in the A cell range were intrasomally recorded and labeled. Of these, 31 were classified as touch, 34 as temperature, and 26 as mechanical nociceptive neurons. The touch neurons were further classified as tactile (TAC, n = 13) and vibrotactile (VIB, n = 18) neurons, and the temperature neurons as thermosensitive (T, n = 15) and thermomechanosensitive (TM, n = 19) neurons according to their response characteristics. The mechanical nociceptive neurons were subdivided into fast-conducting nociceptive (FCN, n = 14) and slow-conducting nociceptive (SCN, n = 12) neurons based on their morphological characteristics and conduction velocities.

The touch neurons had peripheral conduction velocities at A-β range of 24.2–56.8 m/s and the temperature neurons at the A-δ range of 8.4–18.6 m/s. No difference in conduction velocity was found between the TAC and the VIB neurons or between the T and the TM neurons (P > .5 for both). Peripheral conduction velocity of the fast-conducting mechanical nociceptive neurons was 8.2–16.3 m/s, and that of the slow-conducting mechanical nociceptive neurons was 2.4–6.4 m/s. Both of them fell into the A-δ range. No cold neurons and A-δ low-threshold mechanoreceptive afferents were encountered in the present study.

Examples of response characteristics of the classified neurons are shown in Figure 1. Among the sampled cells only the temperature neurons showed background discharges with highly regular intervals between spikes. In general, a tonic response to heat or mechanical stimulation of the receptive field of the thermosensitive and thermomechanosensitive neurons was recorded based on the background discharges (Fig. 1A,B). These two subtypes of temperature neurons exhibited a similar response pattern to heat stimuli. The thermo-mechanosensitive neurons responded to von Frey hair stimuli with thresholds of less than 10 mg, whereas all the thermosensitive neurons failed to respond to such stimuli even if the pressure intensity was increased to more than 1 g. Both of the tactile and vibrotactile neurons responded to light touch of their receptive field with thresholds of 10 mg or less. The tactile neurons exhibited a tonic response containing highly dense spikes in responding to a touch stimulus (Fig. 1C), whereas the vibrotactile neurons showed only one spike in responding to a touch stimulus. All the vibrotactile neurons could respond in the ratio of 1 to 1 to a series of vibrating stimuli with a maximum frequency of 300 Hz (Fig. 1D). The fast-conducting nociceptive and the slow-conducting nociceptive neurons had similar response characteristics; i.e., both of them responded to pricking with thresholds of more

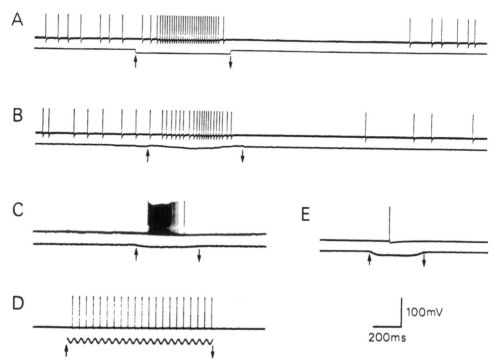

Fig. 1. Response characteristics recorded intrasomally from the trigeminal ganglion neurons. For all the subfigures, the upper trace is the response to a natural stimulus and the lower trace indicates the periods of stimulation (↑, on; ↓, off). **A:** Response to heat stimulation of a thermosensitive neuron. **B:** Response to mechanical stimulation of a thermo-mechanosensitive neuron that responded also to heat stimulus (not shown) with a pattern similar to the thermosensitive neuron. **C:** Response to touch stimulation of a tactile neuron. **D:** Response to vibrating stimulation of a vibrotactile neuron. **E:** Response to pricking stimulation of a fast-conducting A–δ mechanical nociceptive neuron (the slow-conducting A–δ mechanical nociceptive neurons had similar response characteristics, not shown). Stimulus in A was delivered by a laser generating unit, in B, C, and E by a strain gauge, and in D by a vibrating device. Note that only the temperature neurons showed the background discharges. Calibration applies to all the subfigures.

than 5 g, and usually only one spike could be elicited by a stimulus (Fig. 1E). Sensitization to repeated natural stimulation was observed for some of these neurons.

Morphological characteristics

The morphological data of the present study were obtained from the labeled neurons whose soma and stem, peripheral, and central axons as well as axonal bifurcation were well visualized in the trigeminal ganglion.

Soma area. The tactile and vibrotactile neurons had usually a large and oval soma (Fig. 2A,B), whereas the somata of the thermosensitive, thermo-mechanosensitive, fast-conducting nociceptive, and slow-conducting nociceptive neurons were smaller and tended to be round (Fig. 2C–F). The mean areas of the somata with standard deviations and cell numbers of each neuron type are given in Table 1, which was compared in all pairs, and the conclusions are given in Table 2. No significant difference in soma area between the tactile and vibrotactile neurons or

between the arbitrary pair of the thermosensitive, thermo-mechanosensitive, fast-conducting, and slow-conducting nociceptive neurons was observed. However, the soma area of the tactile or vibrotactile neurons was significantly larger than those of other neuron types (Table 2). The distribution of peak area of soma of the tactile and vibrotactile neurons was different from that of the thermosensitive, thermo-mechanosensitive, fast-conducting nociceptive, and slow-conducting nociceptive neurons, but there was a great overlap in the ranges of soma area (Fig. 3). As shown in Figure 2, the soma of neurons gave rise straightforwardly to a simple stem axon that was immediately away from the soma. No glomerular structure in which the initial segment of the stem axon winds repeatedly around its neuronal soma (Spencer et al., 1973) was observed for all labeled neurons in this study.

Axon diameter. Configurations of the HRP-labeled stem, peripheral, and central axons of each type neurons are shown in Figure 4. The stem, peripheral, and central axons

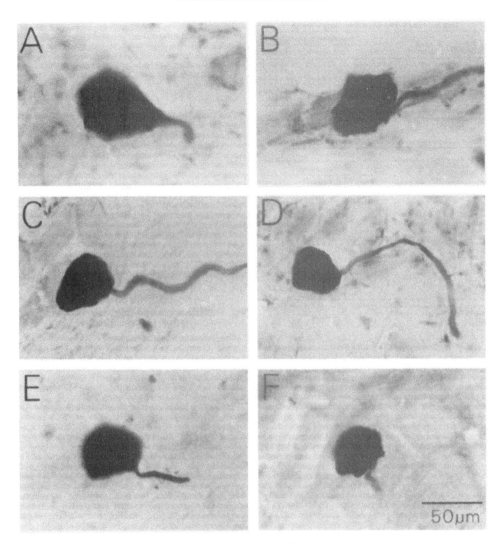

Fig. 2. Photomicrographs of somata and stem axons taken from the horizontal sections of intrasomally HRP-labeled tactile neuron (**A**), vibrotactile neuron (**B**), thermosensitive neuron (**C**), thermo-mechanosensitive neuron (**D**), fast-conducting A–δ mechanical nociceptive neuron (**E**), and slow-conducting A–δ mechanical nociceptive neuron (**F**). Note that somata in A and B are larger and tend to be oval, and those in C–F are smaller and round. Calibration in F applies to all subfigures.

of the tactile and vibrotactile neurons were large, myelinated, and equal in diameter (Fig. 4A,B). The thermosensitive, thermo-mechanosensitive, and fast-conducting nociceptive neurons had a set of myelinated axons of medium size, but the central axon was significantly thinner than the stem and peripheral axons (Fig. 4C–H). The slow-conducting nociceptive neurons had thin myelinated stem and peripheral axons and an unmyelinated central axon (Fig. 4I,J).

Table 1 shows the means of axon diameters of the stem, peripheral, and central axons of each type of neurons, which were compared in all pairs, and the conclusions are given in Table 2. There was no significant difference in axon diameter between the tactile and vibrotactile neurons or

TABLE 1. Soma and Axon Size of the Labeled Neurons[1]

Cell type	Soma area (μm²)	Axon diameter (μm)		
		Stem	Peripheral	Central
VIB	1,568.7 ± 285.6 (18)	6.5 ± 1.4 (18)	6.5 ± 1.3 (18)	6.3 ± 1.5 (18)
TAC	1,529.9 ± 223.5 (13)	6.7 ± 1.5 (13)	6.7 ± 1.4 (13)	6 6 ± 1.6 (13)
T	1,246.1 ± 303.0 (15)	3.8 ± 0.3 (15)*	3.8 ± 0.4 (15)*	2 8 ± 0.4 (15)
TM	1,267.1 ± 326.0 (19)	3.9 ± 0.2 (19)*	3.9 ± 0.2 (19)*	2.9 ± 0 6 (19)
FCN	1,323.9 ± 280.7 (14)	3.8 ± 0.3 (14)*	3.8 ± 0.3 (14)*	2.3 ± 0.5 (14)
SCN	1,196.9 ± 307.2 (12)	2.8 ± 0.4 (12)*	2.7 ± 0.4 (12)*	1 1 ± 0.3 (12)

[1]Values are means ± SD; Numerals in parentheses indicate the No. of neurons
*Significant differences ($P < .001$) comparing the central axons in the same cell type.

TABLE 2. Comparisons of Soma and Axon Size of the Labeled Neurons*

	Probability			
		Axon diameter		
Measurement	Soma area	Stem	Peripheral	Central
VIB:TAC	−	−	−	−
VIB:T	+ +	+ + +	+ + +	+ + +
VIB:TM	+ +	+ + +	+ + +	+ + +
VIB:FCN	+	+ + +	+ + +	+ + +
VIB:SCN	+ +	+ + +	+ + +	+ + +
TAC:T	+	+ + +	+ + +	+ + +
TAC:TM	+	+ + +	+ + +	+ + +
TAC:FCN	+	+ + +	+ + +	+ + +
TAC:SCN	+ +	+ − +	+ + +	+ + +
T:TM	−	−	−	−
T:FCN	−	−	−	+ +
T:SCN	−	+ + +	+ + +	+ + +
TM:FCN	−	−	−	− +
TM:SCN	−	+ + +	+ + +	+ + +
FCN:SCN	−	+ + +	+ + +	+ + +

*+ + +, + +, and +, significant differences with $P < .001$, $P < .01$, and $P < .05$,
respectively; −, without significant difference

between the thermosensitive and thermo-mechanosensitive neurons. However, significant differences were observed when the tactile or vibrotactile neurons were compared with all other neuron types and the thermosensitive, thermo-mechanosensitive, or fast-conducting nociceptive neurons were compared with the slow-conducting nociceptive neurons ($P < .001$, for all). No significant difference in the stem and peripheral axon diameters between the thermosensitive, thermo-mechanosensitive, and fast-conducting nociceptive neurons was observed, but the central axon of the fast-conducting nociceptive neurons was significantly thinner than those of the thermosensitive and thermo-mechanosensitive neurons ($P < .01$). Composite histograms of stem, peripheral, and central axon diameters are shown in Figure 5, in which it is demonstrated that the diameters of the stem and peripheral axons of the studied neurons were in the order of the TAC = VIB > T = TM = FCN > SCN neurons and those of the central axons were the TAC = VIB > T = TM > FCN > SCN neurons. In contrast to the case of the soma area, the distribution of the axon diameter of the tactile and vibrotactile neurons was separated from other neuron types, and there was almost no overlap in the axon diameters (Fig. 5).

Axonal bifurcation. Three patterns of axonal bifurcation in the trigeminal ganglion were visualized in this study (Fig. 4), i.e., 1) all axons constricted at the bifurcation point (Fig. 4A–C,F), 2) axons with a triangular expansion at the bifurcation point (Fig. 4D,G,I), and 3) a central axon arising straightforwardly from the constant stem and peripheral axons (Fig. 4E,H,J). The tactile and vibrotactile neurons had only the first type of bifurcation (Fig. 4A,B). The thermosensitive, thermo-mechanosensitive, and fast-conducting nociceptive neurons exhibited all three types (Fig.

4C–H), and the slow-conducting nociceptive neurons lacked the first type of bifurcation. The third type of bifurcation that has little been discussed in the literature was frequently encountered in the thermosensitive, thermo-mechanosensitive, and fast-conducting nociceptive neurons which had peripheral conduction velocities ranging from 8.2 to 18.6 m/s.

Correlations

The relationships of soma area to the peripheral axon diameter and conduction velocity and of the peripheral axon diameter to the conduction velocity were examined for the individual functional groups and for the whole of labeled neurons. The conclusions of all examined relations with correlation coefficients and neuronal numbers are given in Table 3, and the plots of the relations are presented in Figures 6–8. A positive linear correlation between the peripheral axon diameter and the peripheral conduction velocity existed within the touch neurons (Fig. 6A), the temperature neurons (Fig. 6B), the nociceptive neurons (Fig. 6C), and the overall labeled neurons (Fig. 6D). Significant correlations of the soma area to the peripheral axon diameter and to the peripheral conduction velocity were observed if all labeled neurons were considered together (Figs. 7D, 8D), but no correlations were found within the individual functional groups (Figs. 7A–C, 8A–C).

DISCUSSION

Distinct morphology of identified functional cell types

The present investigation showed that the neurons in the trigeminal ganglia vary in soma size, axon diameter, and axonal bifurcation pattern, and what is more, our data demonstrated for the first time that distinct morphological characteristics of the soma, axon, and axonal bifurcation in the trigeminal ganglia do exist for the touch, temperature, and mechanical nociceptive neurons.

The neurons labeled in the present study had soma sizes belonging to large and median cells in the trigeminal ganglion neurons of crotaline snakes (cf., Terashima and Liang, 1991). Although statistical differences in soma area at the nuclear level between the touch neuron and the other functional cell types were observed (see Table 2) and an impression that the cells with an extremely large soma could be classified as touch neurons and those which had a small soma were devoid of touch neurons may be drawn from the present data, the great overlap in the range of soma areas of touch, temperature, and mechanical nociceptive neurons (see Fig. 3) made it untrustworthy to suggest the sensory modalities from the soma size of the given neurons. Therefore, we concluded that the soma size of the trigeminal ganglion neurons is of little value in predicting their sensory modalities.

Compared to the soma size the axonal morphology appears to be much more closely related to the sensory modalities in this study. Despite some of the soma being smaller, the touch neuron always had a set of thick axons whose diameters were all over 5 μm. In contrast, even if some of the temperature neurons and the nociceptive neurons had a larger soma, they always possessed a set of thinner axons with a diameter less than 4.5 μm. Moreover, the stem, peripheral, and central axons of the touch neurons were equal in diameter, but the central axons of the temperature neurons and the mechanical nociceptive neu-

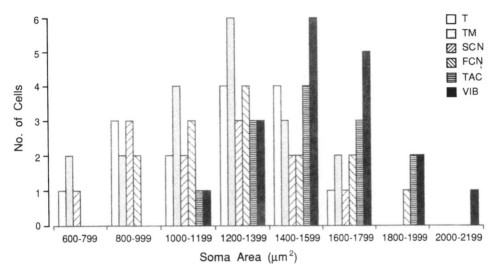

Fig. 3. Composite histogram of soma areas of intrasonally HRP-labeled tactile (TAC), vibrotactile (VIB), thermosensitive (T), thermo-mechanosensitive (TM), fast-conducting A–δ mechanical nociceptive (FCN), and slow-conducting A–δ mechanical nociceptive (SCN) neu-rons. Note that the distribution of soma areas of the T, TM, FCN, and SCN neurons are different from that of the TAC and VIB neurons, but there is a great overlap in the range of 1,000–1,999 μm².

rons were invariably thinner than the stem and peripheral axons. For the temperature and the mechanical nociceptive neurons, the diameters of the central axons were also different. The slow-conducting nociceptive neurons had an unmyelinated and thus the thinnest central axons among the labeled neurons. The fast-conducting nociceptive neurons had a myelinated central axon, but it was significantly thinner than that of the thermosensitive and thermo-mechanosensitive neurons in spite of the fact that they had a similar soma size and peripheral conduction velocity. To a certain extent the axonal bifurcation patterns were also related to the sensory modalities. Although three types of axonal bifurcation were observed in this study, the touch neurons exhibited only one of them, i.e., the stem, periph-eral, and central axons constricted at the bifurcation point. In contrast, the slow-conducting nociceptive neurons just lacked such type of bifurcation and exhibited all the rest, i.e., a triangular expansion existed at the bifurcation point or a central axon arising straightforwardly from the con-stant stem and peripheral axons. Thus, the axonal bifurca-tion pattern appears to be a reliable criterion for distinguish-ing the touch neurons from the slow-conducting nociceptive neurons. However, this criterion cannot be applied to the thermosensitive and thermo-mechanosensitive neurons and the fast-conducting nociceptive neurons because they shared all three types of axonal bifurcation.

Taking all the present data into account, it is possible to predict the sensory modality of a given neuron in the trigeminal ganglion by knowing its morphology, i.e., a cell that has a large and oval soma and a set of large myelinated stem, peripheral, and central axons which are equal in size and are of a constriction at the bifurcation point is likely a touch neuron; a cell that has a median and round soma with thin myelinated stem and peripheral axons and an unmyelin-ated central axon appears to be a slow-conducting mechani-cal nociceptive neuron; and a cell that has a median and round soma and a set of thinner myelinated axons with a myelinated central axon that is thinner than the stem and peripheral axons might be a temperature or a fast-conducting mechanical nociceptive neuron.

The morphological homogeneity in the ganglion between the sub-functional types of the touch neuron and of the temperature neuron implies that distinct morphological characteristics at the receptor level may exist for the sub-functional types because the different sensory modali-ties were usually considered to be generated based on the morphological differences of the receptors (for a review, see Munger, 1971). Moreover, the morphological heterogeneity in the ganglion (present results) and the brainstem (Liang and Terashima, 1993) of the primary mechanical nocicep-tive neurons suggests that the primary neurons which are of the same sensory modality may engender different physiological effects via their peculiar morphological charac-teristics beyond the receptor

Relative size of central and peripheral axons

There has been debate as to whether there is a size difference between central and peripheral axons of sensory ganglion neurons with myelinated axons. Earlier reports stated that the central and peripheral axons of the ganglion cells having myelinated fibers are of equal size (Gasser, 1955; Ha, 1970; Lieberman, 1976; Ochs et al., 1978), but some recent studies yielded data to the contrary. Suh and his co-workers (1984) stated that the mean cross-sectional area of the myelinated central axons of rat dorsal root ganglion cells was 50% less than the myelinated peripheral axons; thus they suggested that the central axons of the

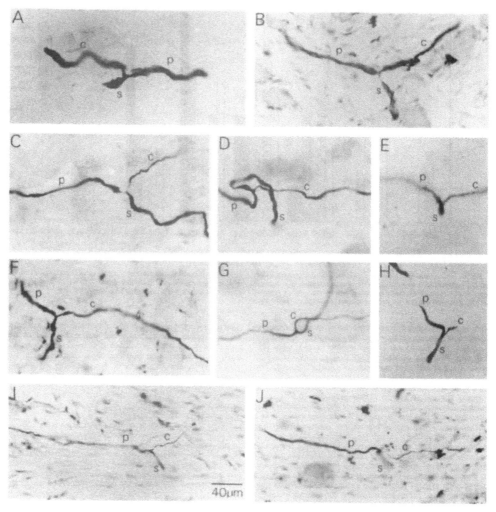

Fig. 4. Photomicrographs of stem (s), peripheral (p), and central (c) axons, and bifurcations taken from the horizontal sections of intra-somally HRP-labeled tactile neuron **(A)**, vibrotactile neuron **(B)**, thermosensitive neuron **(C)**, thermo-mechanosensitive neuron **(D,E)**, fast-conducting A-δ mechanical nociceptive neuron **(F–H)**, and slow-conducting A-δ mechanical nociceptive **(I,J)**. Note that there are three types of bifurcation—i.e., with a constriction of three axons at the bifurcation point (A–C,F), with a triangular expansion at the bifurca-tion point (D,G,I), and with a central axon arising straightforwardly from the constant stem and peripheral axons (E,H,J)—and that the stem, peripheral, and central axons in A and B are thick and equal in size. The axons in C–H are medium in size, and the central axons are thinner than the stem and peripheral axons. The axons in I and J are thin, and the central axons are unmyelinated. Calibration in I applies to all the subfigures.

ganglion cells are thinner than their peripheral axons. In cat, Lee et al. (1986) reported that the mean diameter of the central axons of A cells defined on the basis of impulse conduction velocity was smaller than the peripheral axons. However, reviewing concerned references and taking the present results into account, it is clear that a yes-or-no answer to this question would be an oversimplification, because 1) it was well documented that for the ganglion cell having a large myelinated axon or fast conduction velocity the central and peripheral axons were of equal size (Ha,

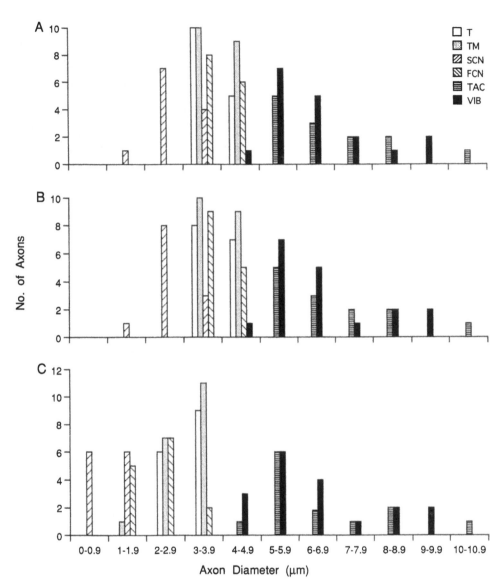

Fig. 5. Composite histograms of stem axon diameters (**A**), peripheral axon diameters (**B**), and central axon diameters (**C**) of intrasomally HRP-labeled tactile (TAC), vibrotactile (VIB), thermosensitive (T), thermo-mechanosensitive (TM), fast-conducting A–δ mechanical nociceptive (FCN), and slow-conducting A–δ mechanical nociceptive (SCN) neurons. Note that the distributions of stem and peripheral axon diameters are different between the TAC and VIB neurons, the T, TM, and FCN neurons, and the SCN neurons (TAC = VIB > T = TM = FCN > SCN), and that the distributions of central axon diameters are different between the TAC and VIB neurons, the T and TM neurons, the FCN neurons, and the SCN neurons (TAC = VIB > T = TM > FCN > SCN).

TABLE 3. Correlations[1]

Measurement	N	r	Regression line
Touch neurons			
PCV/DPA	31	0.960	PCV = 6.68 × DPA − 8.58
PCV/SA	31	0.340	Not significant
SA/DPA	31	0.307	Not significant
Temperature neurons			
PCV/DPA	34	0.801	PCV = 8.51 × DPA − 18.16
PCV/SA	34	0.017	Not significant
SA/DPA	34	0.005	Not significant
Nociceptive neurons			
PCV/DPA	26	0.890	PCV = 6.17 × DPA − 11.91
PCV/SA	26	0.069	Not significant
AS/DPA	26	0.140	Not significant
Whole labeled neurons			
PCV/DPA	91	0.978	PCV = 7.73 × DPA − 16.35
PCV/SA	91	0.517	PCV = 0.02 × SA − 9.34
SA/DPA	91	0.523	SA = 100.09 × DPA + 897.76

[1]PCV, peripheral conduction velocity; DPA, diameter of peripheral axon; SA, soma area.

1970; Kirkwood and Sears, 1982; Rindos et al., 1984; see also this paper); 2) for the ganglion cells, which had a smaller myelinated axon or slow conduction velocity, the central axons have been proved to be thinner than the peripheral axons (Hoheisel and Mense, 1987; Terashima and Liang, 1991; Liang and Terashima, 1993; see also this paper); and 3) a less mean cross-sectional area (Suh et al., 1984) or a less mean diameter (Lee et al., 1986) of the myelinated central axons could not rule out the existence of the neurons which have equal size of peripheral and central axons. An appropriate conclusion on this question appears to be that the ganglion cells with thick myelinated axons usually have equal-sized peripheral and central axons, those with median myelinated axons frequently have a central axon that is thinner than the peripheral axon, and

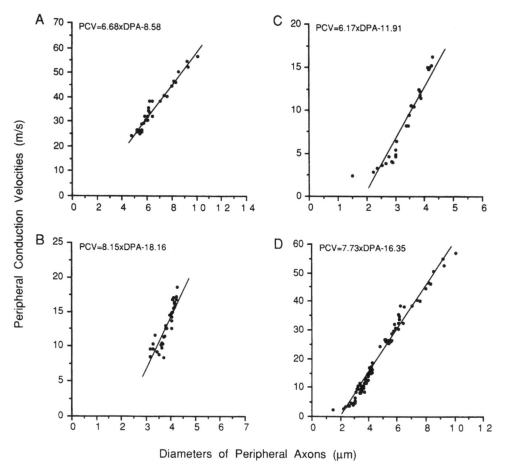

Fig. 6 Plots of the relations between diameters of the peripheral axons (DPA) and peripheral conduction velocities (PCV). **A:** Plot for touch neurons. **B:** Plot for temperature neurons. **C:** Plot for nociceptive neurons. **D:** Plot for all labeled neurons. The oblique line in each subfigure indicates significant linear regression. Formula for each line is given on the upper side of the corresponding subfigure.

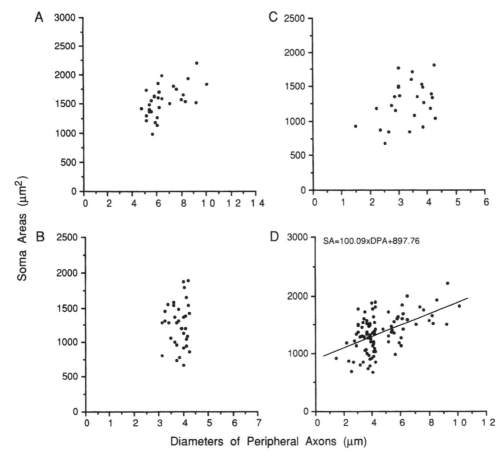

Fig. 7. Plots of the relations between diameters of the peripheral axons (DPA) and soma areas (SA). **A:** Plot for touch neurons. **B:** Plot for temperature neurons. **C:** Plot for nociceptive neurons. **D:** Plot for all labeled neurons. Note that significant linear regression as indicated by the oblique line was found only in D. Formula for the line is given on the upper side of the subfigure.

those with thin myelinated axons often have a much thinner or even unmyelinated central axon.

The present morphological observation that the temperature neurons had central axons thinner than their peripheral axons, the mechanical nociceptive neurons had central axons much thinner than their peripheral axons, and the touch neurons had equal-sized myelinated central and peripheral axons corresponds well to the physiological observation made by Traub and Mendell (1988): They reported that in the cat dorsal root ganglion A–δ neurons exhibited a decrease in conduction velocity and that such a decrease was much greater for A–δ high-threshold mechanoreceptive neurons, but there was no decrease in conduction velocity for A–α and A–β neurons. Since it is generally accepted that the conduction velocity varies with the axon size (Hursh, 1939; Boyd and Kalu, 1979), it is reasonable to

suppose that a decrease of the central axon size would result in a decrease in conduction velocity in the ganglia. Thus, it appears that a decrease of conduction velocity in the trigeminal ganglion may occur on the temperature neurons and much greater on the nociceptive neurons, but not on the touch neurons.

Relationships of conduction velocity, axon diameter, and soma area

Studies on the relationship between conduction velocity and axon size of the myelinated fibers have revealed a strong positive correlation between these two parameters. However, the scaling factors, converting axon diameter in microns into conduction velocity in meters per second, vary. In the cat Hursh (1939) stated a scaling factor of 6.0 for the

Y.-F. LIANG *et al.*

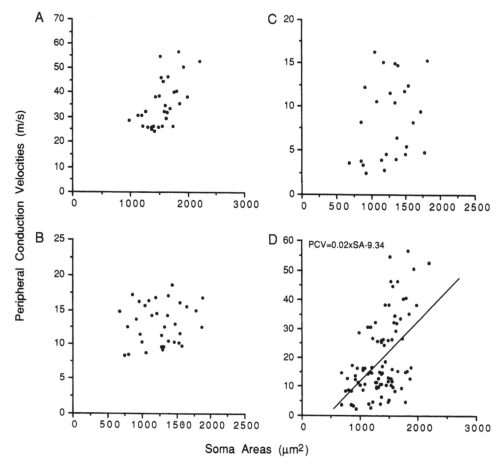

Fig. 8. Plots of the relations between soma areas (SA) and peripheral conduction velocities (PCV). **A:** Plot for touch neurons. **B:** Plot for
temperature neurons. **C:** Plot for nociceptive neurons. **D:** Plot for all
labeled neurons. Note that significant linear regression as indicated by
the oblique line was found only in D. Formula for the line is given on the
upper side of the subfigure.

largest myelinated axons, but Boyd and Kalu (1979) reported a smaller scaling factor of 5.7 for group I neuronal
axons and of 4.6 for group II and III neuronal axons. A
positive linear correlation between conduction velocity and
axon size was also found in our study (see Fig. 6), but the
scaling factor, 3.1, was much lower than those reported by
Hursh (1939) and Boyd and Kalu (1979). Arbuthnott and
his co-workers (1980) reported that the thickness of the
myelin sheath was linearly related to axon diameter for
group II and III neurons. If such correlation is applicable to
our material, then we suggested that the scaling factors are
not only different from the cell types defined by conduction
velocity but are also different according to species, being
smaller in the snake than in the cat.

As to the relationship of soma size to axon size or
conduction velocity of the ganglion cells, a generally accepted concept is that large cells give rise to thick axons,
and thus have fast conduction velocity, and that small cells
have thin axons and slow conduction velocity. Recently, by
injection of a marker into the cells defined by conduction
velocity a linear relation between soma size to their peripheral axon diameter and/or conduction velocity was observed
in the dorsal root ganglion when the A and C cells were
considered together (Yoshida and Matsuda, 1979; Cameron
et al., 1986; Hoheisel and Mense, 1987). Harper and
Lawson (1985) reported a loose correlation for the fast-
conducting A cells and a closer correlation for A–δ cells in
rat. Lee and his co-workers (1986) found also a loose

correlation for A cells with conduction velocity of more than 2.5 m/s in rat. Although we observed a linear correlation of soma area to axon diameter and conduction velocity for the overall labeled neurons which had axons conducting in the A range, our observation is somewhat different from that of Harper and Lawson (1985) but more consistent with that of Hoheisel and Mense (1987), i.e., no significant correlation was observed for A–β or A–δ neurons alone.

It is noteworthy that in all concerned studies, in which a significant correlation of soma size to axon diameter or conduction velocity has been obtained, the scatter of data in the correlation diagram was obviously great. Therefore, we agree with the conclusion that the axonal diameter and conduction velocity of a neuron is not a decisive factor for determining the size of its soma (Mense, 1990). Since similar results were obtained from different animals (cat, Lee et al., 1986; Hoheisel and Mense, 1987; snake, this study) this conclusion appears not to be species specific. Moreover, our data showed that for the cells which conduct in the A range, it is impossible to predict a soma size from its axonal diameter or conduction velocity and vice versa within individual identified functional groups.

LITERATURE CITED

Adams, J.C. (1981) Heavy metal intensification of DAB-based HRP reaction product (letter). J. Histochem. Cytochem. 29:775.

Arbuthnott, E.R., I.A. Boyd, and K.U. Kalu (1980) Ultrastructural dimensions of myelinated peripheral nerve fibers in the cat and their relation to conduction velocity. J. Physiol. (Lond.) 308:125–157.

Boyd, I.A., and K.U. Kalu (1979) Scaling factor relating conduction velocity and diameter for myelinated afferent never fibers in the cat hindlimb. J. Physiol. (Lond.) 289:277–297.

Burgess, P.R., and E.R. Perl (1973) Cutaneous mechanoreceptors and nociceptors. In A. Iggo (ed): Handbook of Sensory Physiology, Somatosensory System. Heidelberg: Springer, vol. 2, pp. 29–78.

Cajal, S.R. (1906) Struktur der sensiblen Ganglien des Menschen und der Tiere. Anat Hefte Zweite Abt 16:177–215.

Cameron, A.A., J.D. Leah, and P.J. Snow (1986) The electrophysiological and morphological characteristics of feline dorsal root ganglion cells. Brain Res. 362:1–6.

Dogiel, A.S. (1898) Zur Frage uber den Bau der Spinal ganglien beim Menschen und bei den Saugetieren. Int. Monatsschr. Anat. Physiol. 15:343.

Gallego, R., and C. Eyzaguirre (1978) Membrane and action potential characteristics of A and C nodose ganglion cells studied in whole ganglia and in tissue slices. J. Neurophysiol. 41:1217–1232.

Gasser, H.S. (1955) Properties of dorsal root unmedullated fibers on both the sides of the ganglion. J. Gen. Physiol. 38:709–728.

Ha, H. (1970) Axonal bifurcation in the dorsal root ganglion of the cat: a light and electron microscopic study. J. Comp. Neurol. 140:227–240.

Harper, A.A., and S.H. Lawson (1985) Conduction velocity is related to morphological cell type in rat dorsal root ganglion neurone. J. Physiol. (Lond.) 359:31–46.

Hatai, S. (1902) Number of size of the spinal ganglion cells and dorsal root fibres in the white rat at different ages. J. Comp. Neurol. 12:107–124.

Hoheisel, U., and S. Mense (1987) Observations on the morphology of axons and somata of slowly conducting dorsal root ganglion cells in the cat. Brain Res. 423:269–278.

Hursh, J.B. (1939) Conduction velocity and diameter of nerve fibers. Am. J. Physiol. 127:131–139.

Kirkwood, P.A., and T.A. Sears (1982) Excitatory post-synaptic potentials from single muscle spindle afferent in external intercostal motoneurons of the cat. J. Physiol. (Lond.) 322:287–341.

Jacquin, M.F., W.E. Renehan, B.G. Klein, R.D. Mooney, and R.W. Rhoades (1986) Functional consequences of neonatal infraorbital nerve section in rat trigeminal ganglion. J. Neurosci. 6(12):3706–3720.

Jacquin, M.F., J.W. Hu, B.J. Sessle, W.E. Renehan, and P.M.E. Waite (1992) Intra-axonal neurobiotin® injection rapidly stains the long-range projections of identified trigeminal primary afferents in vivo: comparisons with HRP and PHA-L. J. Neurosci. Methods 45:71–86.

Lee, K.H., K. Chung, J.M. Chung, and R.E. Coggeshall (1986) Correlation of cell body size, axon size, and signal conduction velocity for individually labeled dorsal root ganglion cell in the cat. J. Comp. Neurol. 234:335–346.

Liang, Y.-F., and S. Terashima (1993) Physiological properties and morphological characteristics of cutaneous and mucosal mechanical nociceptive neurons with A–δ peripheral axons in the trigeminal ganglia of crotaline snakes. J. Comp. Neurol. 328:88–102.

Lieberman, A.R. (1976) Sensory ganglion in the peripheral nerve. In D.N. Landon (ed): The Peripheral Nerve. London: Chapman and Hall, pp. 188–278.

Mense, S. (1990) Structure-function relationships in identified afferent neurons. Anat. Embryol. (Berl.) 181:1–17.

Munger, B.L. (1971) Patterns of organization of peripheral sensory receptors. In W.R. Loewenstein (ed): Handbook of Sensory Physiology, Principles of Receptor Physiology. New York: Springer-Verlag, 1:523–556.

Nagy, I., L. Urban, and C.J. Woolf (1993) Morphological and membrane properties of young rat lumbar and thoracic dorsal root ganglion cells with unmyelinated axons. Brain Res. 609:193–207.

Ochs, S., J. Erdman, R.A. Jersild, Jr., and V. McAdoo (1978) Routing of transported materials in the dorsal root and nerve fiber branches of the dorsal root ganglion. J. Neurobiol. 9:465–481.

Ranson, S.W. (1912) The structure of the spinal ganglia and of the spinal nerves. J. Comp. Neurol. 22:159–175.

Ranson, S.W., and H.K. Davenport (1931) Sensory unmyelinated fibers in the spinal nerves. Am. J. Anat. 48:331–353.

Rindos, A.J., G.E. Loeb, and H. Levitan (1984) Conduction velocity changes along lumbar primary afferent fibers in cats. Exp. Neurol. 86:208–226.

Spencer, P.S., C.S. Raine, and H. Wisniewski (1973) Axon diameter and myelin thickness—unusual relationships in dorsal root ganglion. Anat. Res. 176:225–244.

Suh, Y.S., K. Chung, and R.E. Coggeshall (1984) A study of axonal diameters and areas in lumbosacral root and nerves in the rat. J. Comp. Neurol. 222:473–481.

Terashima, S., and R.C. Goris (1979) Receptive area of primary infrared afferent neurons in crotaline snakes. Neuroscience 4:1137–1144.

Terashima, S., and Y.-F. Liang (1991) Temperature neurons in the crotaline trigeminal ganglia. J. Neurophysiol. 66:623–634.

Terashima, S., and Y.-F. Liang (1993) Modality difference in the physiological properties and morphological characteristics of the trigeminal sensory neurons. Jpn. J. Physiol. 43(Suppl 1):S267–274.

Traub, R.J., and L.M. Mendell (1988) The spinal projection of individual identified A–δ and C fibers. J. Neurophysiol. 59:41–55.

Yoshida, S., and Y. Matsuda (1979) Studies on sensory neurons of the mouse with intracellular-recording and horseradish peroxidase injection techniques. J. Neurophysiol. 42:1134–1145.

Somatosensory and visual correlation in the optic tectum of a python, *Python regius*: a horseradish peroxidase and Golgi study

Sonou Kobayashi*[a], Fumiaki Amemiya[a], Reiji Kishida[b], Richard C. Goris[a], Toyokazu Kusunoki[a], Hironobu Ito[c]

[a]*Department of Anatomy, School of Medicine, Yokohama City University, Yokohama, Japan*
[b]*Department of Anatomy, School of Medicine, Yamaguchi University, Ube, Japan*
[c]*Department of Anatomy, Nippon Medical School, Tokyo, Japan*

Abstract

In snakes with infrared receptors the optic tectum receives infrared input in addition to visual and general somatosensory inputs. In order to observe their tectal termination patterns in ball pythons, *Python regius*, we injected horseradish peroxidase (HRP) into the nucleus of the lateral descending trigeminal tract (LTTD) which mediates infrared information, the optic nerve, and the nucleus of the trigeminal descending tract (TTD) which relays general somatosensory information. Fibers from LTTD were found in layers 5–13 of the contralateral optic tectum, and were especially dense in layers 7a–8. Optic nerve fibers terminated in layers 7a–13 of the contralateral tectum, and mainly in layers 12–13. TTD fibers were few, and could be seen in only the rostral half of the contralateral tectum. These fibers were found in layers 5–7b, but mainly in layers 6–7a. Among various types of neurons stained by the Golgi-Cox method, we focused on six types of neurons whose dendritic arborization overlapped with the distribution of the terminals of these sensory afferents described above. It is possible that these different sensory modalities converge on a single neuron of the various types.

Keywords: Python; Optic tectum; Vision; Infrared sensory; Somatosensory; Correlation; Horseradish peroxidase; Golgi-Cox

1. Introduction

The question of how various types of sensory input influence each other during processing of information in the central nervous system is as yet unsolved. The optic tectum receives several types of sensory input and is one of the largest correlation centers in the central nervous system (Ito et al., 1980, 1981, 1982, 1984, 1992). However, it is not yet known how the optic tectum processes this information.

In snakes with infrared receptors the optic tectum receives infrared input in addition to visual and other general somatosensory inputs (Kishida et al., 1980; Newman et al., 1980; Newman and Hartline, 1982). With electrophysiological recording and marking techniques in the tectum, Kass et al. (1978) demonstrated the existence of bimodal neurons, which are responsive to both visual and infrared stimuli of pit vipers. We have recently demonstrated by electron microscopy that both retinal and infrared inputs converge on a single dendrite in the python optic tectum (Kobayashi et al., 1992). When the left eyeball was enucleated and horseradish peroxidase (HRP) was injected into the left lateral descending trigeminal tract (LTTD) in the same animal, both degenerating and HRP-labeled terminals appeared to make synaptic contacts on a single dendrite in the right tectum. However, we could not identify under the electron microscope what type(s) of neurons received the two different sensory modalities.

In the present study on pythons, we attempted to clarify the following points by means of anterograde labeling of sensory afferents and the Golgi staining of tectal neurons, (1) the arrangement of the different

* Corresponding author, Department of Oral and Maxillofacial Surgery, Yokohama Seamen's Insurance Hospital, Kamadai-cho 137, Hodogaya-ku, Yokohama, Japan.

layers of the optic tectum, which receive fibers from the eyes, pit organs, and general somatosensory receptors, and of the endings of these fibers, (2) the morphology of the neurons receiving this input, and (3) how these three different types of information relate to each other in the optic tectum.

In snakes, input from the eyes passes through the optic nerve to the contralateral optic tectum (Armstrong, 1951; Gruberg et al., 1979; Schroeder, 1981). The nerve supplying the pit organs is the trigeminal nerve, which also serves other somatosensory receptors in the head and neck area. In the medulla, there is a specialized first relay nucleus for infrared input called the LTTD (Molenaar, 1974; Schroeder and Loop, 1976). In pit vipers, input from infrared receptor cells in the pit organ passes through the trigeminal ganglion to the LTTD (Gruberg et al., 1979; Newman et al., 1980), and to the nucleus reticularis caloris, and then terminates in the contralateral optic tectum (Kishida et al., 1980; Newman et al., 1980). In pythons and boas there is no nucleus reticularis caloris, so that input from the infrared receptors passes from the trigeminal ganglion through the LTTD directly to the contralateral optic tectum (Molenaar, 1974; Molenaar and Fizaan-Oostveen, 1980; Newman et al., 1980).

Input from the general somatosensory receptors of the head and neck area passes through the trigeminal ganglion to the nucleus of the descending trigeminal tract (TTD) to the contralateral optic tectum or the thalamus (Welker et al., 1983).

The retinorecipient layers in a rattlesnake tectum have been reported (Schroeder, 1981). There is also a report which describes the projection of infrared sensory neurons to the tectum in a python (Molenaar and Fizaan-Oostveen, 1980), using degeneration techniques. The TTD-tectal pathway has been described using retrograde HRP techniques (Welker et al., 1983), but there are no reports of anterograde labeling of the terminal layers of fibers projecting from the TTD, and there are no reports on the terminal layers of fibers projecting from the optic nerve, the LTTD, and the TTD in one single species of snake and using the same method.

In the present study, therefore, we injected HRP into the optic nerve, the LTTD, and the TTD of pythons and observed the terminal layers in the optic tectum. We also observed the dendritic arborization patterns of neurons in Golgi-Cox stained specimens to learn what kind of neurons could be receiving the three types of information. Combining these two lines of study we will discuss how different sensory inputs are integrated in the optic tectum in the snake.

2. Materials and methods

2.1. Materials

Twelve ball pythons, *Python regius*, about 100 cm in

length and 1 kg in weight were used. Three were given an injection of HRP into the LTTD, three were given an injection of HRP into the TTD, three were injected with HRP in the optic nerve, and three were used for Golgi-Cox staining.

2.2. Anesthesia

All surgical procedures were performed under general anesthesia administered by a flow-through method (Goris and Terashima, 1973). The animals were premedicated with atropine sulfate, 0.01 mg/kg, by intramuscular injection. General anesthesia was slowly induced with ketamine hydrochloride, 1 mg/kg, by intramuscular injection. The animals were then intubated and anesthesia was maintained by a flow of 0.1–0.2% fluothane in moist oxygen from a modified small-animal respirator. The gas entered through a tube inserted in the glottis and exited through large-bore hypodermic needles inserted into the posterior right and left lung sacks. We monitored the general condition of the animals with an electrocardiograph machine.

2.3. HRP injection into the optic nerve

Under general anesthesia, an incision was made in the muscles around the eyeball and the opening of the optic canal was exposed. The optic nerve was freed from the central artery and vein and the surrounding tissues, and exposed. A glass micropipette was used to inject 50% HRP (Toyobo Co., Japan) in 0.3 μl of distilled water containing 2.5% L-a-lysophosphatidylcholine (Sigma) into the optic nerve.

2.4. HRP injection into the LTTD and TTD

Because we have studied fiber connections of the infrared system using the HRP methods for a long time, we have had considerable experience in injecting a small amount of HRP into the LTTD and TTD in both boid (*P. molurus* and *Boa constrictor*, Kishida et al., 1983) and crotaline (*Agkistrodon blomhoffi*, Kishida et al., 1980) snakes. On the basis of this experience we were able to inject HRP into each nucleus discretely, without cross-diffusion. For the LTTD we ablated the temporal muscles from the skull under general anesthesia and retracted them. Next we severed the quadrate from the skull and opened a hole with a dental engine and a round burr through the inner ear to expose the medulla. The meninges were slit with a hypodermic needle. Then a glass micropipette on a manipulator was used to inject slowly (over a period of 30 min) 30% HRP in 1 μl of distilled water containing 2.5% L-a-lysophosphatidylcholine into the LTTD.

For the TTD we exposed the medulla by the same method, and exposed the lower edge of the superior sagittal sinus. Then we injected HRP into the TTD by the method described above, taking care that none diffused to the LTTD (Fig. 1).

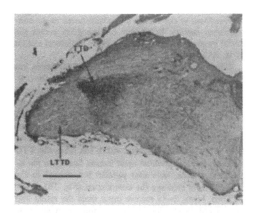

Fig. 1. Frontal section of the medulla oblongata of the python. HRP was restricted to the TTD without diffusing to the LTTD. Scale bar = 500 μm.

2.5. Preparation of HRP specimens

After the operation each python was given two intra-muscular injections of 4 mg/kg of amikacin sulfate at an interval of 2 days and kept alive for 7–13 days at a temperature of 25°C and 50% humidity with free access to drinking water. Then the animal was anesthetized with fluothane and perfused through the heart with heparinized (5 units/ml) 0.9% saline, followed by perfusion with 1% paraformaldehyde and 4% glutaraldehyde in 0.1 M phosphate buffer at pH 7.4. The brains were then removed, postfixed in the same fresh fixative for 2–3 h, and stored in phosphate buffer containing 30% sucrose until they sank. Frontal frozen serial sections were cut at 40 μm and mounted alternately on two series of gelatinized glass slides. The sections were processed for HRP with 3,3'-diaminobenzidine (DAB) method (Adams, 1981), and alternate sections were counter-stained with cresyl violet.

2.6. Specimens for Golgi-Cox study

Ninety micrometer serial frontal sections were cut and stained by the Golgi-Cox method (Ramón-Moliner, 1970).

3. Results

We followed Ramón in assigning numbers (Ramón, 1896) to the layers of the tectum opticum.

3.1. HRP study

Labeled optic nerve fibers ran from the retina across the optic chiasm to the contralateral optic tectum. We found labeled fibers only in the contralateral tectum, and none in the ipsilateral tectum. A small number of optic fibers terminated in the lateral geniculate body of contralateral side, and an even smaller number of fibers in the lateral geniculate body of ipsilateral side. In the

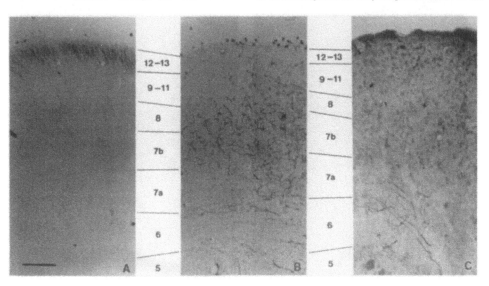

Fig. 2. Frontal sections of the python tectum. A: fibers from the optic nerve were found in layers 7a–13, and were especially dense in layers 12–13. B: fibers from the LTTD were found in layers 5–13, and were especially dense in layers 7a–8. C: TTD fibers were few, and could be seen in only the rostral half of the tectum. These fibers were found in layers 5–7b, but mainly in layers 6–7a. Scale bar = 100 μm.

tectum, labeled fibers from the optic nerve were found in layers 7a–13. Labeled fibers were few in layers 7a and 7b, but were especially dense in layers 12–13 (Fig. 2A). Labeled fibers in the layers were thin, had few branches, and ran mostly parallel in a single direction from the surface of the tectum to the deeper layers, where they were distributed at a comparatively low density.

Labeled fibers from the LTTD ran first ventrally in the medulla, then medially to cross the midline. From there they coursed in a bundle along the outer edge of the medulla, passing through the tegmentum and then dorsally and medially again inside the tectum. Fibers from the LTTD were found in layers 5–13 of the optic tectum, and were especially dense in layers 7a–8 (Fig. 2B). LTTD fibers were thicker and more numerous than the optic nerve fibers, branched frequently, ran in various directions at random, and were densely distri-

buted. Most of the LTTD fibers terminated in the contralateral tectum, but a few continued on across the tectal commissure to terminate in the ipsilateral tectum (Fig. 3A).

Fibers from the TTD paralleled the LTTD fibers, but ran medially to them. Most of the fibers ran to the contralateral thalamus, but a few ran to the contralateral tectum. TTD fibers were rather few compared with LTTD fibers, and could be seen in only the rostral half of the optic tectum. These fibers were found in layers 5–7b, but mainly in layers 6–7a (Fig. 2C). There were also a few ipsilateral fibers, which ran from the TTD through the contralateral optic tectum directly to the ipsilateral tectum, here also parallel but ventral to the LTTD fibers crossing the commissure (Fig. 3B).

Thus, layers 8–13 receive input from the retina and the LTTD, layers 7a–7b receive input from all three

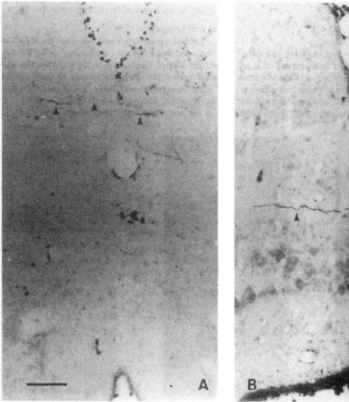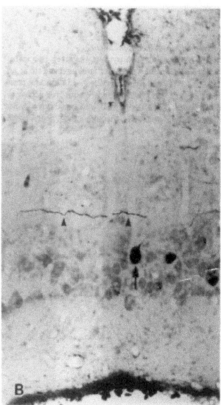

Fig. 3. Frontal sections of the tectum showing HRP-labeled fibers (arrowheads). A, LTTD fibers; B, TTD fibers. A soma of the trigeminal mesencephalic nucleus has also been labeled here (arrow). Scale bar = 100 μm.

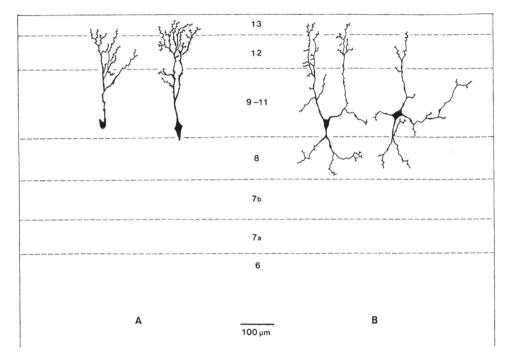

Fig. 4. Drawing of neuron types from Golgi-Cox specimens. A, type 1; B, type 2.

sources (retina, LTTD, and TTD), and layers 5–6 receive input from the LTTD and the TTD.

3.2. Golgi-Cox study

Neurons of the optic tectum stained by the Golgi-Cox method were extremely variable, but we focused on the following six types because the dendritic arborization of these neurons overlapped with the distribution of terminals from the retina, LTTD, or TTD.

Type 1 (Fig. 4A). Neurons of this type were unipolar with medium-sized (about 30×20 μm) pyriform or fusiform somata located in layers 9–11. They had one apical dendrite with 3–4 main branches, which extended into layers 12–13, where they branched. The branchlets were characteristically spiny. The apical dendrites were about 200 μm in length.

Type 2 (Fig. 4B). These neurons were multipolar with medium-sized (about 30×20 μm) pyriform or fusiform somata located in layers 9–11. Four to five main dendrites (150–300 μm in length) extended upward to layers 12–13 and downward to layers 8–11. The branches were studded with a few spines and varicosities.

Type 3 (Fig. 5A). These neurons were also multipolar, with characteristically large (about 40×20 μm) triangular somata located in layer 8. They had a few main dendrites, which extended upward for more than 300 μm to layers 8–13, and downward for about 200 μm to layers 7a–7b. In layers 8–13, the dendrites branched repeatedly giving them a 'bushy' appearance.

Type 4 (Fig. 5B). Neurons of this type had small (about 20×15 μm) pyriform somata located in layer 7b. A few long (250–300 μm) dendrites extended upward as far as layers 8–11, and a few short (100–200 μm) downward dendrites extended to layers 6–7a. These dendrites had sparse spines and a few varicosities.

Type 5 (Fig. 6A). These were bipolar neurons with small (about 20×15 μm) oval somata located in layers 6 and 7a. The dendrites were short (100–200 μm) and extended either horizontally or vertically with a few branches.

Type 6 (Fig. 6B). These were bipolar neurons with medium-sized (about 30×20 μm) oval somata and characterized by a long (400–500 μm) slender upward dendrite. The somata were located in layer 6, and their

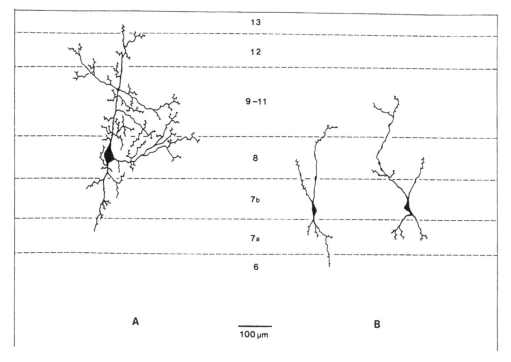

Fig. 5. Drawing of neuron types from Golgi-Cox specimens. A, type 3; B, type 4.

long dendrites extended upward as far as layers 9–11. A few short downward dendrites extended to layers 5–6. A few spines were seen on these dendritic branches.

4. Discussion

This work has demonstrated the terminal layers of fibers from the optic nerve, the LTTD, and the TTD in the python optic tectum for the first time using the same method for each fiber type.

Schroeder (1981) reported that retinal fibers terminate in the stratum zonale, the stratum opticum, and the stratum griseum et fibrosum centrale in the rattlesnake (*Crotalus viridis*). These strata correspond to layers 7b–14 of the present study. Thus her observations coincide almost completely with ours, although the fibers were much more numerous than in our material. This could be the result of using different species.

Using degeneration techniques, Molenaar and Fizaan-Oostveen (1980) demonstrated the projection of infrared sensory neurons to the tectum opticum in pythons (*P. reticulatus*). They reported that LTTD fibers terminate in the stratum griseum centrale, which

corresponds to layers 6–7b of our study. In our study, fibers from the LTTD were more widely distributed in layers 5–13, and had many branches. This distribution is broader than has been previously reported, and there is also a considerable overlap with the terminal distribution of visual modality. This undoubtedly reflects the importance of the infrared information for pythons. Goris and Terashima (1973) reported a similar high degree of development of the terminal layers in the optic tectum of a nocturnal pit viper, as compared to the tectum of a nocturnal snake not possessing the infrared sense. Masai and Sato (1969) also reported that the input from the receptors, most vital to the life of a given species, influences most strongly the architecture of the brain of that species.

Welker et al. (1983) were the first to demonstrate the TTD-tectal pathway. They used retrograde HRP techniques, and reported results similar to ours, i.e., that fibers projecting from the TTD were more numerous on the contralateral side, with some projections on the ipsilateral side.

Neurons of the optic tectum stained by the Golgi-Cox method were of many types (Northcutt, 1984), but we

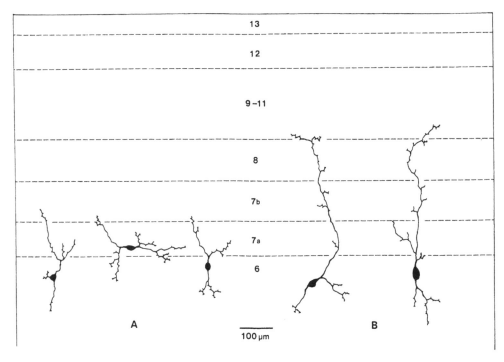

Fig. 6. Drawing of neuron types from Golgi-Cox specimens. A, type 5; B, type 6.

focused on six types, since dendritic arborization of these neurons overlapped with projection layers from the retina, LTTD, or TTD. Type 1 seemed to receive fibers from the optic nerve, because the somata were located in layers 8–11, and their dendrites extended to layers 12–13, where they branched. Type 2 seemed to be receiving fibers from both the retina and the LTTD, because the somata were located in layers 9–11, and their dendrites extended to layers 12–13 and 9–11, and branched. Type 3 seemed to be receiving fibers from three sources, the retina, the LTTD, and the TTD, because the somata were located in layer 8, and their dendrites extended to layers 7a–13 with many branches. Type 4 seemed to be receiving LTTD and TTD inputs, because the somata were located in layer 7b, with the apical dendrite extending to layers 6–7a and 8–11. The somata of type 5 neurons were located in layers 6 and 7a, with less extensive arborization than the other types, so that it is highly possible that they received input from only one of the two modes (LTTD and TTD), although we cannot exclude the possibility that at least some are bimodal. Type 6 seemed to be receiving fibers from both the LTTD and the TTD, because the somata were

located in layer 6, and their dendrites extended from layers 5–9, and branched.

With electrophysiological recording and marking techniques, Kass et al. (1978) were able to demonstrate localization of three types of neurons in the tectum: visual, infrared, and bimodal visual-infrared. They reported that neurons responsive to infrared stimulation are seen mainly in layers 7a and 7b in the pit viper tectum, whereas reaction to visible light is seen in layers 7b–13. Bimodal neurons are found predominantly in layer 7b. Kobayashi et al. (1992) demonstrated infrared and visual synapses on one and the same dendrite in the python (*P. regius*) tectum using HRP and degeneration techniques. These dendrites were located in layers 7b and 8, and probably belonged to type 2 or type 3 neurons (see description above).

In the tectum, retinal fibers enter the superficial layers, LTTD fibers the middle layers, and TTD fibers the deep layers. The visual, infrared and somatic representations are topographically organized in these layers of termination. In spite of this laminar segregation of different sensory modalities, the retinotopic, infrared-topic and somatotopic maps in the tectum are roughly in register with each other (Terahsima and Goris, 1975; Hartline et al., 1978; Stein, 1984). That is, nasal visual fields and rostral body sectors are represented in the lateral tectum, and temporal visual fields and caudal body sectors are in the medial tectum in *Iguana*. In the same way, upper visual fields and upper body sectors are represented in the rostral tectum, and lower visual fields and lower body sectors in the caudal tectum (Stein, 1984). In the present work as well, this kind of mapping could be inferred from the fact that TTD fibers were present only in the rostral half of the tectum. Since TTD fibers carry information from the head and neck region, this shows that the point-to-point mapping of the TTD fibers is the same as that of LTTD and retinal fibers, in which forward-facing (rostral) receptive fields are mapped onto the rostral part of the tectum, and proceed backwards as a stimulus moves toward the tail of the snake (Hartline et al., 1978; Stein, 1984).

According to Hartline et al. (1978), bimodal neurons form local connections without regard to establishing spatial correspondence between the visual and infrared modalities. Tectal receptive neurons classed as 'or', 'and', and 'enhanced' exhibit excitatory interactions of infrared and visual inputs. Two 'depressed' classes of cells respond to only one of the two stimuli and show inhibitory interactions when both inputs are presented together. Such physiological diversity in bimodal neurons undoubtedly depends on the morphology of the connections bearing input to the neuron; e.g., whether the cell receives infrared input on one dendrite and visual input on another, or receives the two modalities on the same dendrite, or receives one modality on a small proportion of its dendrites and the other on a larger proportion of the dendrites, etc. The physiological responses could also change according to whether one is recording from a neuron that is processing information independently or from one that is a single link in a processing chain, as discussed above.

Acknowledgements

We wish to thank Dr. Yoshihide Tanaka and Dr. Satoshi Hasegawa for technical assistance, and Mrs. Chikako Usami and Mrs. Miki Kobayashi for their excellent help in preparing the histological materials and illustrations. A part of this research was supported by Grant 01440019 for scientific research from the Ministry of Education of Japan to H.I..

References

Adams, J.C. (1981) Heavy metal intensification of DAB-based reaction product. J. Histochem. Cytochem., 29: 775.

Armstrong, J.A. (1951) An experimental study of the visual pathways in a snake (*Natrix natrix*). J. Anat., 85: 275–288.

De Cock Buning, T.J., Goris, R.C. and Terashima, S. (1981) The role of thermosensitivity in the feeding behavior of the pit viper, *Agkistrodon blomhoffi brevicaudus*. Jpn. J. Herpetol., 9: 7–27.

Goris, R.C. and Terashima, S. (1973) Central response to infra-red stimulation of the pit receptor in a crotaline snake, *Trimeresurus flavoviridis*. J. Exp. Biol., 58: 59–76.

Gruberg, E.R., Kicliter, E., Newman, E.A., Kass, L. and Hartline, P. H. (1979) Connections of the tectum of the rattlesnake, *Crotalus viridis*: an HRP study. J. Comp. Neurol., 188: 31–41.

Hartline, P.H., Kass, L. and Loop, M.S. (1978) Meaning of modalities in the optic tectum: infrared and visual integration in rattlesnakes. Science, 199: 1225–1229.

Ito, H., Butler, A.B. and Ebbesson, S.O.E. (1980) An ultrastructural study of the normal synaptic organization of the optic tectum and the degenerating tectal afferents from retina, telencephalon and contralateral tectum in a teleost, *Holocentrus rufus*. J. Comp. Neurol., 191: 639–659.

Ito, H., Tanaka, H., Sakamoto, N. and Morita, Y. (1981) Isthmic afferent neurons identified by the retrograde HRP method in a teleost. Brain Research, 207: 163–169.

Ito, H., Sakamoto, N., and Takatsuji, T. (1982) Cytoarchitecture, fiber connections, and ultrastructure of the nucleus isthmi in a teleost (*Navodon modestus*), with a special reference to the degenerating isthmic afferents from the optic tectum and nucleus pretectalis. J. Comp. Neurol., 205: 299–311.

Ito, H., Vanegas, H., Murakami, T. and Morita, Y. (1984) Diameters and terminal patterns of retinofugal axons in their target areas. An HRP study in two teleosts (*Sebastiscus* and *Navodon*). J. Comp. Neurol., 230: 179–197.

Ito, H., Yoshimoto, M., Uchiyama, H. and Negishi, K. (1992) Changes in retinal projections and ganglion cell morphology after unilateral enucleation in common carps. Brain Behav. Evol., 40: 197–208.

Kass, L., Hartline, P.H. and Loop, M.S. (1978) Anatomical and physiological localization of visual and infrared cell layers in tectum of pit vipers. J. Comp. Neurol., 182: 811–820.

Kishida, R., Amemiya, F., Kusunoki, T. and Terashima, S. (1980) A new tectal afferent nucleus of the infrared sensory system in the medulla oblongata of crotaline snakes. Brain Res., 195: 271–279.

Kishida, R., Buning, T.C. and Dubbeldam, J.L. (1983) Primary vagal nerve projections to the lateral descending trigeminal nucleus in boidae (*Python molurus* and *Boa constrictor*). Brain Res., 263: 132–136.

Kobayashi, S., Kishida, R., Goris, R.C., Yoshimoto, M. and Ito, H. (1992) Visual and infrared input to the same dendrite in the tectum opticum of the python, *Python regius*: electron-microscopic evidence. Brain Res., 597: 350–352.

Masai, H. and Sato, Y. (1969) The brain of archaic fishes. Zool. Magazine, 78: 187–195.

Molenaar, G.J. (1974) An additional trigeminal system in certain snakes possessing infrared receptors. Brain Res., 78: 340–344.

Molenaar, G.J. and Fizaan-Oostveen, J.L.F.P. (1980) Ascending projections from the lateral descending and common sensory trigeminal nuclei in python. J. Comp. Neurol., 189: 555–572.

Newman, E.A., Gruberg, E.R. and Hartline, P.H. (1980) The infrared trigemino-tectal pathway in the rattlesnake and in the python. J. Comp. Neurol., 191: 465–477.

Newman, E.A. and Hartline, P.H. (1982) The infrared 'vision' of snakes. Sci. Am., 246: 98–107.

Northcutt, R.G. (1984) Anatomical organization of the optic tectum in reptiles. In: H. Vanegas (Ed.), Comparative Neurology of the Optic Tectum, Plenum Press, New York and London, pp. 548–600.

Ramón-Moliner, E. (1970) The Golgi-Cox technique. In: W.G.H. Nauta and S.O.E. Ebbesson (Eds.), Contemporary Research Methods in Neuroanatomy, Springer-Verlag, New York, pp. 32–55.

Ramón, P. (1896) Estructura del encefalo del cameleon. Rev. Trimest. Micrograf., 1: 46–82.

Schroeder, D.M. (1981) Retinal afferents and efferents of an infrared sensitive snake, *Crotalus viridis*, J. Comp. Neurol., 170: 29–42.

Schroeder, D.M. and Loop, M.S. (1976) Trigeminal projections in snakes possessing infrared sensitivity. J. Comp. Neurol., 169: 1–14.

Stein, B.E. (1984) Multimodal representation in the superior colliculus and optic tectum. In: H. Vanegas (Ed.), Comparative Neurology of the Optic Tectum, Plenum Press, New York and London, pp. 819–841.

Terashima, S. and Goris, R.C. (1975) Tectal organization of pit viper infrared reception. Brain Res., 83: 490–494.

Welker, E., Hoogland, P.V., and Lohman, A.H.M. (1983) Tectal connections in *Python reticulatus*. J. Comp. Neurol., 220: 347–354.

THE SURFACE ARCHITECTURE OF SNAKE INFRARED RECEPTOR ORGANS

Fumiaki Amemiya[1], Richard C. Goris[1], Yoshiki Masuda[2], Reiji Kishida[3], Yoshitoshi Atobe[1], Norihisa Ishii[4] and Toyokazu Kusunoki[1]

[1]Department of Anatomy, Yokohama City University School of Medicine, Fukuura, Kanazawaku, Yokohama 236, [2]Department of Biology, Kawasaki Medical School, Kurashiki 701-01, [3]Department of Anatomy, Yamaguchi University School of Medicine, Ube, Yamaguchi 755, and [4]Department of Dermatology, Yokohama City University School of Medicine, Yokohama 236, Japan

ABSTRACT

The surface of the epithelium in snake infrared receptor organs is covered with a characteristic array of tiny pores that is different from any other surface structure in squamate reptiles. The measurements and density of the pores differ slightly according to family and species, but the array is characteristic and immediately recognizable. In boids without pits, the array covers the entire surface of each scale that contains infrared receptors. In boids with pits, the array covers the fundus of each receptor pit organ. In crotaline pit organs the array is present on both the outer and inner surfaces of the receptor-containing membrane, and on the epithelium of the wall of the inner chamber. This inner chamber wall is sculpted into a tight array of large and small domed structures, on the surface of which the pore array appears. We speculate that the array of domes in the crotaline pit organ functions as a light trap to prevent infrared rays that penetrate into the inner chamber from being reflected back onto the receptors in the pit membrane. On the other hand, the array of pores, present in all species, appears to reflect away and diffuse visible radiation that might have enough energy to heat-stimulate the receptors and interfere with the target stimulus, i.e., infrared radiation.

Two groups of snakes possess infrared receptors: those of the family Boidae (the boids), and those of the subfamily Crotalinae (the crotalines) of the family Viperidae. The boids have the receptors in the labial scales, some without specialized structures, and others in the fundus of specialized labial pits. The crotalines, on the other hand, have the receptors in a thin membrane suspended between inner and outer chambers of a pair of pits in the loreal region. The three different types are illustrated in Fig. 1.

While doing a series of comparative anatomical studies on the structure of the infrared receptor organs, we were struck by certain surface features that seemed to be possessed in common by all species having these organs, and that had never before been reported in the literature. Since these surface features were common to all species with

infrared reception, and were found only in regions associated with the infrared receptors, we surmised that they played a major role in the receptor function. As a first step in clarifying this role, we did a scanning and transmission electron microscope study of all surface areas associated with these receptors.

MATERIALS AND METHODS

As representative of the evolutionarily advanced crotaline snakes, we used 3 *Agkistrodon blomhoffii*, a common pit viper of Japan, Korea, and the eastern regions of China. For the boids with pits, we used 4 *Python regius*, the ball python, and 1 *Python molurus*, the Burmese python. For a representative of the boids that do not have labial pits, we used 1 *Boa constrictor*, the common boa con-

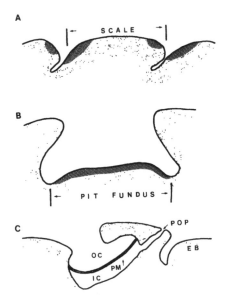

A

SCALE

B

PIT FUNDUS

C

POP

OC

EB

PM

IC

Fig. 1 Schematic cross sections of the various types of snake infrared receptor organs. Hatching indicates the location of receptor nerve terminals, based on SDH staining. A: Boids without pits. The receptors are grouped immediately beneath the keratinized surface of the labial scales. B: Boids with pits. The receptors are located beneath the keratinized surface of the fundus of a number of pits in the labial scales. C: Crotaline snakes. The receptors are contained in the pit membrane (PM) suspended between the outer (OC) and inner (IC) chambers of a single pit in the loreal region. Note the presence of the preocular pore (POP). EB, eyeball

strictor of Central and South America. As a control, representing the snakes that do not have infrared reception, we used 1 specimen of the common colubrid *Elaphe obsoleta quadrivittata*, the yellow rat snake of the eastern United States.

All boid and crotaline specimens were anesthetized with halothane, and perfused through the right aortic arch with heparinized saline solution containing an additional anesthetic, tricaine methanosulfonate, followed by 2% paraformaldehyde and 2.5% glutaraldehyde in 0.1 M phosphate buffer at pH 7.4. The appropriate tissues were dissected from the head, postfixed in 2% osmium tetroxide, and dehydrated with ethanol.

Specimens destined for transmission electron

microscopy were embedded in a mixture of Epon and Araldite or in Luft's Epon mixture. One μm semithin sections stained with toluidine blue were used for light microscopy and for selecting typical areas for ultrathin sections. Ultrathin sections of the chosen areas were then stained with uranyl acetate and lead citrate.

Specimens for scanning electron microscopy were critical-point dried and sputter-coated with gold-palladium or platinum-palladium alloy.

The control animal was anesthetized with halothane and one upper labial scale and one parietal scale were dissected away. The snake was then allowed to recover. As further controls, we cut away a parietal and some dorsal scales from *B. constrictor*, and the scales outside one of the pits of *P. molurus*. The dissected tissues were fixed and prepared for scanning electron microscopy as described above.

In crotaline snakes it has long been known that the infrared receptors are entirely contained in the membrane suspended between the two chambers of the loreal pit (see 5, for review). The receptors in the membrane contain large numbers of mitochondria, and can be visualized by staining for the succinate dehydrogenase (SDH) present in the mitochondria (2). To locate the receptors in the boid snakes, we also used SDH staining. The method and results have been published elsewhere (1).

RESULTS

Boids without Pits

In *Boa constrictor*, our SDH staining confirmed the work of von Düring (9), who described the presence of infrared receptors in this species in supralabial scales 8, 9, and 10, and in the 3 subocular scales (6) above these. Anticipating the presence of receptors in most, if not all, of the labial scales, we stained right supralabials 3 and 13, and found receptors localized in the rostral upper corner and caudal lower corner of each scale (Fig. 2A). We did not examine the subocular scales because the labials alone provided more than enough material for the work planned.

The entire surface of scales containing infrared receptors presented a pitted appearance, due to the presence of an array of microscopic pores (Fig. 2, B–D). The pores were roundish or elliptical, 0.3–0.5 μm in diameter at the mouth and 0.15–0.25 μm in depth, occupying only the outermost portion of the cornified layer. They were more or less crater-like in cross section (Fig. 2B), and spaced at inter-

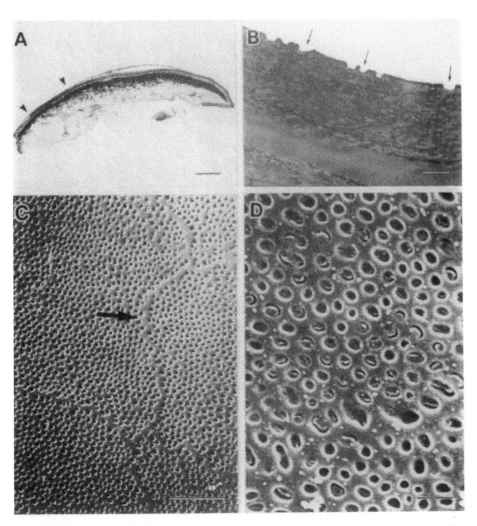

Fig. 2 Surface structure of the pit organs of boids without pits (here, *B. constrictor*). A: Partial section of an infrared receptive scale. SDH staining of the receptors shows them grouped in the epidermis in only a part of the scale (between arrowheads). The rest of the epidermis is unstained. The black staining in the dermis is due to the presence of melanin. Bar = 200 μm. B: A transmission electron microscope (TEM) image shows the microscopic pores of the surface in cross section (arrows). Note the crater-like shape. Bar = 1 μm. C: A low magnification scanning electron microscope (SEM) image of an infrared receptive scale. The ridge marked with an arrow is the raised border between oberhauchten cells. Bar = 1 μm. D: A high magnification SEM image of the same scale. Bar = 1 μm

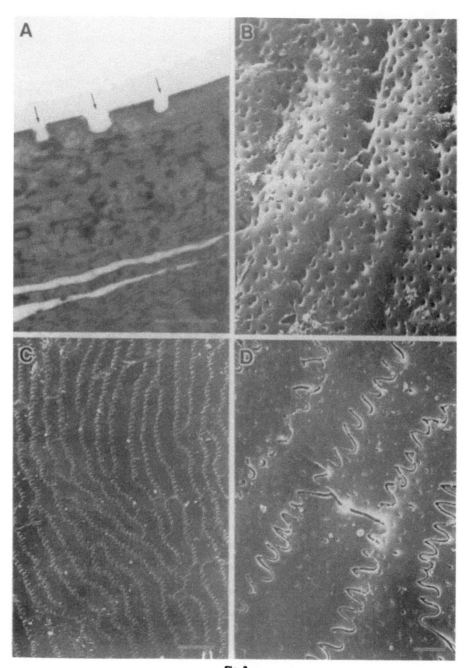

Fig. 3

vals of 0.15–0.3 μm, with an average density of 3.34/μm^2 (Fig. 2, C and D). Scales not containing infrared receptors had a quite different surface architecture (Fig. 3, C and D). The pores were not confined to the areas of each scale containing receptors, but extended over the entire surface of the scale.

Boids with Pits

Both species examined had arrays of pores similar to those described above for *B. constrictor*, but different in size, shape, and density. The pores were cylindrical in cross section (Fig. 3A), 0.1–0.25 μm in diameter and 0.1–0.25 μm in depth, spaced at intervals of 0.15–0.45 μm, with an average density of 5.45/μm^2 (Fig. 3B). The pores were present only at the bottom of the receptor pits, in the area where SDH staining revealed the presence of receptors.

Crotaline Snakes

As mentioned in Materials and Methods, the crotaline infrared organs consist of an inner and outer chamber separated by a thin (15 μm in *Trimeresurus flavoviridis*, see 8) membrane which contains the receptor terminal nerve masses (*ibidem*). We found an array of pores on both the outer and the inner surfaces of this membrane (Fig. 4). We also found specialized structures on the surface of the wall of the inner chamber (Fig. 5).

The outer surface of the pit membrane was divided into mostly pentagonal, sometimes hexagonal, oberhauchten cells with well-defined, raised borders (Fig. 4B). The cells measured 15–30 μm at the widest point between opposing sides. All cells were covered with an array of pores similar to those of the boids with pits described above. The pores were crater-like in cross section (Fig. 4A),

Fig. 3 Surface structure of the pit fundus in boids with pits. A: A TEM image of the outer keratinized layer (here, *P. regius*). The pores (arrows) are more cylindrical than in boids without pits. Magnification is greater here than in Fig. 2B, although the pores themselves are slightly smaller. Bar = 0.5 μm. B: A SEM image of the surface of the pit fundus (here and in C and D, *P. molurus*). Note the characteristic pored image, in contrast to the sculpted image in C and D. Bar = 1 μm. C: A SEM image of the scale surface outside the pit. Bar = 5 μm. D: C at higher magnification. Some pores can be seen here also, but they are smaller, less regular in shape, and fewer than the infrared organ pores in B. Bar = 1 μm

0.25–0.5 μm in diameter at the mouth and 0.2–0.25 μm in depth, and spaced at intervals of 0.15–0.4 μm, with an average density of 3.66/μm^2.

The inner surface of the pit membrane was also divided into pentagonal or sometimes hexagonal oberhauchten cells. In this case the cells were larger than the outer ones, measuring 20–40 μm at the widest point between opposing sides. This inner surface was also covered with an array of pores that were crater-like in cross section. They were 0.06–0.25 μm in diameter at the mouth and spaced at intervals of 0.25–0.5 μm, with a density of 1.97/μm^2. In contrast to the outer surface, the pentagonal or hexagonal cells had a slightly raised center where the density of the pores was reduced to about 1.50/μm^2 (Fig. 4C). These central pores were also smaller than the others, being 0.06–0.1 μm in diameter. We were unable to measure the depth of the pores on the inner surface of the membrane due to lack of suitable transmission electron microscope specimens.

The surface of the inner chamber presented the most specialized, complex picture. The entire surface was covered with domed structures of fairly regular size (Fig. 5). There were 2 classes of these domes: large (15–20 μm in diameter at the base and 8–11 μm from base to apex), and small (2–6 μm in diameter at the base and 1–3 μm from base to apex) (Fig. 5A). The large domes were spaced so closely that they were nearly touching, at an average density of 3,160/mm^2. The small domes packed the spaces occurring between the round bases of the large domes. There were 3–4 small domes in each space, depending on their size. At the base they appeared to be in direct contact with each other and with the large domes, and their density per unit area was 13,400/mm^2. Thus the wall of the inner chamber was completely covered with domes, with no open spaces whatever (Fig. 5, B–D).

A striking feature of the domes was that their surfaces, both those of the large domes and those of the small domes, were covered with an array of pores that were similar in shape, size, and density to those of the outer surface: crater-like, 0.25–0.5 μm in diameter at the mouth, and spaced at intervals of 0.25–0.5 μm, with a density of 2.41/μm^2 (Fig. 5C).

The dome-covered epithelium was not confined only to the wall of the inner chamber, but continued on through the pre-ocular pore (2) into the portion of the orbit immediately caudal to the pit (Fig. 6A). However, as one proceeded away from

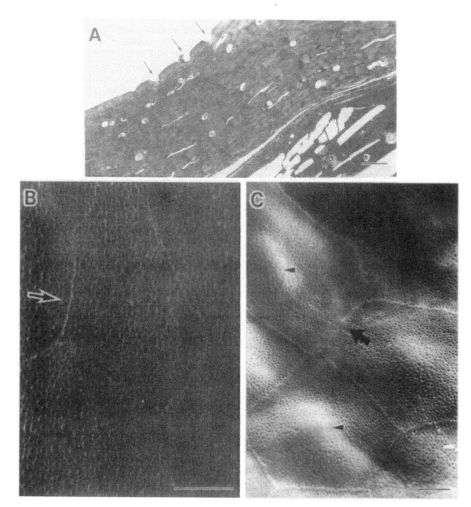

Fig. 4 Surface of the pit membrane in crotaline snakes (here and in Figs. 5 and 6, *A. blomhoffii*). A: A TEM image of the outer keratinized layer of the pit membrane. Arrows point to the pores. Bar = 1 μm. B: A SEM image of the outer surface of the pit membrane. Arrows here and in C point to the raised borders of the oberhauchten cells. Bar = 5 μm. C: A SEM image of the inner surface of the pit membrane. The image is quite similar to that of B, with the exception of the raised centers of the oberhauchten cells (arrowheads). Bar = 5 μm

the pit itself, the specialization of the epithelium became less and less apparent: the small domes eventually disappeared entirely, the large domes became more or less distorted, and the pores covering their surface became smaller, shallower, and less regular in form and distribution (Fig. 6B).

Controls

All controls, i.e. scales from snakes without infrared reception, and boid and crotaline scales from areas not related to infrared reception, showed features common to squamate reptiles, as reviewed aptly by Landmann (4) (Fig. 3, C and D). But none of the controls showed the characteristic array of rounded pores that we have described above.

DISCUSSION

Significance of the Pores

The scale surface in squamate reptiles generally presents a sculpted, pitted appearance (see 4, for review). However, the array of rounded pores that we have described in this paper can be seen only in snakes that possess infrared receptors; and in these snakes only in structures immediately associated with the receptors themselves, but nowhere else. For example, in the crotaline *A. blomhoffii*, not even the scales at the immediate edge of the pit organs have the array of rounded pores, but instead show the sculpted appearance common to other body scales, with scattered, small, irregular pores (Fig. 3, C and D). In the boid without pits, *B. constrictor*, the pore array is seen in places where there are no infrared receptors, but it is still confined within the borders of the scales that contain infrared receptors. Other scales show the common sculpted appearance.

The pore arrays are seen even in the tissues behind the receptors in crotaline snakes, i.e., in the domed epidermis of the inner chamber of the pit organs. They even appear on the domed surface of this epidermis as it continues on through the preocular pore to the orbit of the eye, although outside the confines of the pit they become shallower and less regular in appearance (Fig. 6B).

This fact, i.e., that representative species of the family and subfamily possessing infrared reception, although of widely separated evolutionary stock (cf. 1), all possess a similar array of pores covering or backing up the infrared receptors, can mean only one thing. That is, the pores serve in some way to facilitate infrared reception. We speculate that the pores serve as a form of filter, reflecting away the shorter wavelengths of light in the visual spectrum, while allowing the longer infrared wavelengths free passage through the keratin to the receptors below. This could be necessary because some of the longer visual wavelengths could have enough energy to raise the temperature of the receptors and cause a nerve potential that would diminish the sharpness of the infrared 'image'. As a matter of fact, Terashima and Goris (7) were able to do very precise electrophysiological measurements of receptor performance using a visible laser (He-Ne, 632.8 nm wavelength) that had sufficient energy. This speculation is strengthened by simple observation of the pit organs with the naked eye. Compared to surrounding skin structures the pit membranes of crotalines and the fundus of boid pits are extremely reflective of visual light, to the extent that when photographing at close range with a flash or spotlight, one must pay attention to the angle at which the light strikes the pit organ in order to avoid unseemly glare.

To obtain this effect of rejecting visible light to enhance the receptivity of infrared wavelengths, the precise measurements of the array of pores do not seem to be critical. As a matter of fact, there is a certain amount of difference among species in the diameter, depth, distribution, and density per unit area of the pores, as illustrated in Table 1. Shape and all measurements are roughly equal in the labial scales of *B. constrictor*, and in the outer surface of the pit membrane and the domed wall of the inner chamber of the pit organ in *A. blomhoffii*. In *P. regius* and *P. molurus*, on the other hand, the pores are cylindrical instead of crater-like, smaller and shallower, and distributed at greater density. The reason for this difference is not readily apparent. However, the similarity is such that the array is immediately recognizable at suitable magnification, so that it is possible to predict the presence or absence of infrared receptors by observation of the skin surface with, e.g., a scanning electron microscope.

In boids without pits the pores were distributed over the entire surface of scales containing receptors, including sections of the scale where there were no receptors. This is not surprising if we consider the scale in its entirety as the infrared receptor organ. Thus the specialization of the epidermis continues over the entire surface of the organ, in order to cut down unwanted reflections.

Significance of the Domes

The domed structure of the inner chamber epitheli-

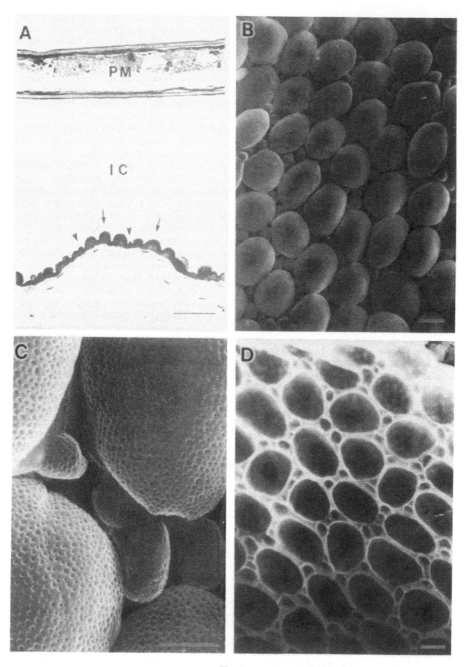

Fig. 5

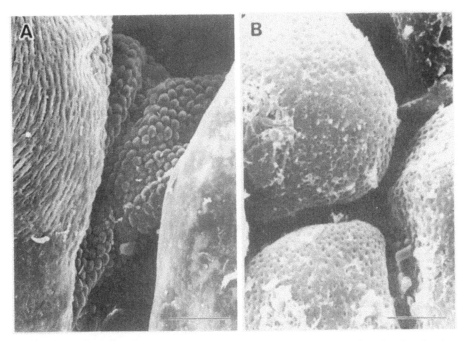

Fig. 6 SEM images of the inner wall of the preocular pore. A: The wall shows a domed surface, but the domes are more irregular than inside the pit, and the small domes are lacking. Bar = 100 μm. B: A at higher magnification. The domes have pores on their surface, but the pores appear less regular than inside the pit. The absence of small domes is also very apparent. Bar = 5 μm

Fig. 5 Surface structures on the wall of the inner chamber of the crotaline pit organ. A: A light micrograph of a semithin cross section through the pit organ stained with toluidine blue. Arrows point to the large domes, arrowheads to the small domes. PM, pit membrane; IC, inner chamber. Bar = 50 μm. B: A SEM image of the surface of the inner wall. The large and small domes can be clearly distinguished. Bar = 10 μm. C: B at higher magnification, showing the characteristic pores on the surface of both the large and small domes. Bar = 5 μm. D: A SEM image of the underside of the keratinized surface of the domes of B. The measurements of dome diameters were made on images such as this. Bar = 10 μm

um of crotaline snakes undoubtedly also functions to enhance the sharpness of infrared 'imaging', but in a slightly different way from the pore array. In crotalines, evolution has increased the sensitivity of infrared reception by an order or two of magnitude by drastically reducing the heat capacity of the receptor-bearing structures. That is, by segregating the receptors from the rest of the body and suspending them in an extremely thin membrane, it has become possible for a given nervous response to be elicited by a far smaller amount of radiant energy than is necessary to obtain the same response in boid pits where the receptors are in contact with the rest of the head tissues. This increase in sensitivity, however, carries with it a heavy cost. The pit membrane is now so thin that it will not absorb all the infrared radiation that impinges on it. A large amount penetrates through the membrane (3), with the danger that it will be

Table 1   *Dimensions of the Surface Structures of Snake Infrared Receptor Organs*

		Boids without pits	Boids with pits	Crotaline snakes
Pores				outer/inner
	Diameter	0.3 –0.5	0.1 –0.15	0.25–0.5/0.06–0.25
	Depth	0.15–0.25	0.05–0.1	0.2 –0.25/no data
	Interval	0.15–0.3	0.15–0.45	0.15–0.4/0.25–0.5
	Density	$3.34/\mu m^2$	$5.45/\mu m^2$	$3.66/1.97/\mu m^2$
	Shape	crater-like	cylindrical	crater-like/crater-like
Domes				
Large	Diameter			15.0–20.0
	Height			8.0–11.0
	Density			$3,160/mm^2$
Small	Diameter			2.0–6.0
	Height			1.0–3.0
	Density			$13,400/mm^2$

All values in μm except where specified. Outer/inner = pit membrane outer surface/pit membrane inner surface

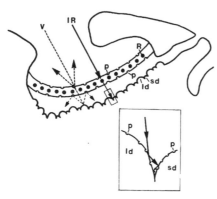

Fig. 7   Schematic summary of the function of the surface structures as typified in a crotaline snake infrared receptor organ. Infrared rays (IR) pass through the surface layer of the pit membrane and stimulate the receptors (R). Some of the IR rays pass entirely through the membrane, but are absorbed by the light trap formed by the large (ld) and small (sd) domes (enlarged view at bottom). Visual light (V) is largely reflected by the pores (p) at the surface of the pit membrane, but some passes through. This light is in turn diffused by the suface pores of the inner chamber domes. Any visual light that is reflected back to the pit membrane is further diffused by the pores on the inner surface of the membrane. The final result is an excellent signal-to-noise ratio for the infrared stimulus.

reflected back from the head tissues to re-stimulate the receptors, causing a large amount of amorphous, background 'noise'.

To obviate this difficulty, the crotalines have evolved the domed wall of the inner chamber to serve as a trap for visual and infrared radiation, analogous to the 'light trap' (10) antireflective coating used on lens hoods and other such optical instruments. In other words, the combination of large and small domes forms a sort of maze in regard to impinging infrared rays. Once they enter the maze they bounce back and forth among the domes interminably but are unable to exit. This prevents an infrared stimulus which has excited a receptor from bouncing straight back from the opposite direction to cancel out the signal it has just generated. The tiny pores on the surface of the domes probably serve the same function as they do on the outer surface of the pit membrane. That is, they effectively disperse rays in the visual spectrum that have passed through the membrane, improving the signal-to-noise ratio. The same function can be attributed to the pores on the inner surface of the pit membrane. They most likely serve to further disperse any random reflections from the wall of the inner chamber, improving the signal-to-noise ratio even further.

As recorded in Results, the domed structure could be seen also inside the preocular pore and in the epithelium of the orbit. Since there is no clear demarcation between the epithelium of the inner chamber and that of the preocular pore and the orbit, this does not seem strange: the epithelium of the pore and the orbit is simply a continuation of that of the pit. However, since neither the inside of the pore nor the orbit epithelium are exposed to infrared radiation, the domes of these two areas probably have no direct connection with the functioning of the pit. In fact, the irregularity of the domes in these extralimital areas, and the lack of small domes between them, strengthens the hypothesis that the regular array of large and small domes inside the inner chamber does indeed function as a light trap to heighten signal-to-noise ratio.

Conclusion

The physical configuration of the surfaces of snake infrared receptor organs serves to make the organs eminently suitable for their purpose. I.e., they are highly selective in regard to the portion of the electromagnetic spectrum that they have been evolved to 'see'. Among the various forms of these organs, the pit organs of the crotaline snakes, with their combination of pores and domes, can be said to be the most effective infrared sensors known.

We wish to thank in particular Mrs Chikako Usami and Mrs Miki Kobayashi for their technical support.

REFERENCES

1. AMEMIYA F., GORIS R. C., ATOBE Y., ISHII N. and KUSUNOKI T. (1995) The ultrastructure of infrared receptors in a boid snake, *Python regius:* evidence for periodic regeneration of the terminals. *Anim. Eye Res.* (in press)
2. GORIS R. C., KADOTA T. and KISHIDA R. (1989) Innervation of snake pit organ membranes mapped by receptor terminal succinate dehydrogenase activity. In *Current Herpetology in East Asia* (ed. MATSUI M., HIDAKA T. and GORIS R. C.) Herpetological Society of Japan, Kyoto, pp. 8–16
3. GORIS R. C. and NOMOTO M. (1967) Infrared reception in oriental crotaline snakes. *Comp. Biochem. Physiol.* **23**, 879–892
4. LANDMANN L. (1986) Epidermis and dermis. In *Biology of the Integument, 2, Vertebrates*, IV The Skin of Reptiles (ed. BEREITER-HAHN J., MATOLTSY A. G. and SYLVIA RICHARDS K.) Springer-Verlag, Berlin, pp. 150–187
5. MOLENAAR G. J. (1992) Sensorimotor integration. In *Biology of the Reptilia, Vol. 17, Neurology C.* (ed. GANS C. and ULINSKI P. S.) The University of Chicago Press, Chicago, pp. 367–453
6. PETERS J. A. (1964) *Dictionary of Herpetology*, Hafner Publishing Company, New York
7. TERASHIMA S. and GORIS R. C. (1977) Infrared bulbar units in Crotaline snakes. *Proc. Japan Acad.* **53**, 292–296
8. TERASHIMA S., GORIS R. C. and KATSUKI Y. (1970) Structure of warm fiber terminals in pit membrane of pit vipers. *J. Ultrastruct. Res.* **31**, 494–506
9. VON DÜRING M. (1974) The radiant heat receptor and other tissue receptors in the scales of the upper jaw of Boa constrictor. *Z. Anat. Entwickl. Gesch.* **145**, 299–319
10. WOOD R. W. (1934) *Physical Optics* Third Ed., The MacMillan Co., New York

C mechanical nociceptive neurons in the crotaline trigeminal ganglia

Shin-ichi Terashima*, Yun-Fei Liang

Department of Physiology, University of the Ryukyus School of Medicine, Nishihara-cho, Okinawa 901-01, Japan

Abstract

Using 32 Crotaline snakes, *Trimeresurus flavoviridis*, intrasomal recordings were made from 44 neurons of the trigeminal ganglia in vivo. They were 10 C neurons from 9 snakes and 34 A-delta mechanical nociceptive neurons from 23 snakes. 5 of the 10 C neurons were identified as mechanical nociceptive neurons. The neurons were labeled with iontophoretically injected HRP. Each of the 5 C nociceptive neurons had one receptive field, on which 1 spike was elicited by pricking the skin or mucosa with a pin. They were sensitized after repeated stimulation. The fields were insensitive to thermal stimulation. No background discharge was observed. Average conduction velocity was 0.95 m/s ($\pm$ 0.4 S.D., $n = 5$). Mean resting potential was -62.5 mV ($\pm$ 6.0 S.D., $n = 4$), and mean action potential amplitude was 88.0 mV ($\pm$ 10.9 S.D., $n = 4$). Two somata were successfully visualized with HRP (22 μm × 20 μm, 20 μm × 18 μm). Total lengths of labeled axons were 1260 and 1480 μm peripherally to the edge of the section, and 1810 and 770 μm centrally. Neither of the neurons had branching of the peripheral or central axons in the ganglion.

Key words: C nociceptive neuron; Trigeminal ganglion; Unmyelinated fiber; Intrasomal recording; Intrasomal HRP injection; Small ganglion neuron

Groups of ganglion neurons are usually classified by conduction velocity (CV), and their action potentials (APs) are related both to the specific type [8,9,12,24] and to the modality [2,4,15,17,23,26,29]. Even in the same modality, the action potential may vary among submodalities, as in slow conducting (SC) and fast conducting (FC) mechanical nociceptive neurons [17].

The responses of C nociceptive neurons have been reported [3,6]. They are known to have a T bifurcation in the ganglion, with a thick peripheral axon and a thin central axon (cf. [18,19]), but the morphology and AP parameters of these neurons have remained uncertain.

The purpose of this study was to collect data concerning physiological and morphological properties of C nociceptive neurons, and compare the data with those obtained from A-delta nociceptive neurons in previous experiments [17]. A part of this report was published elsewhere [27,28].

Snakes of the oriental Crotaline species *Trimeresurus*

flavoviridis were used as experimental animals [7]. Anesthesia was induced with halothane; when necessary pancuronium (2 mg/kg i.m.) was used with artificial respiration. For intrasomal recording and horseradish peroxidase (HRP, Sigma type VI) injection glass micropipettes were filled with 4–6% HRP in 0.5 M KCl and 0.05 M Tris-buffer (pH 7.3). A microelectrode was inserted into the trigeminal ganglion, and intrasomal potentials were recorded. For DC recording a conventional electronic system was used. For dV/dt analysis an electronic differentiator (Nihon Kohden, SS-1468 modified) was used. The neurons were stimulated first by pricking their receptive field with a pin (≥ 5 g), and then by electrical stimulation from the superficial or deep branch of the maxillary division of the trigeminal nerve. CV was calculated from the measured distance and the time required for conduction. Temperature response (0–60 °C) was tested with a metal rod (2 mm in diameter) applied to the receptive field, but chemical response was not tested. After recording intrasomal potentials HRP was injected by iontophoresis.

Two or three days after the HRP injection the animal

*Corresponding author. Fax: (81) (98) 895-6569.

was perfused through the heart with a fixative (2% paraformaldehyde and 2.5% glutaraldehyde in 0.1% phosphate buffer, pH 7.4), and the brain and trigeminal ganglion with the nerves were removed. Serial frozen sections were cut at 80 μm, processed for HRP with a modified diaminobenzidine (DAB) method [1], and counterstained with Cresyl violet. The neurons were observed under a light microscope. Our nomenclature follows Molenaar [20,21]. Other details of the methods were published elsewhere [17].

Forty-four neurons were recorded from 32 snakes of which 10 were C neurons and 34 were A-delta nociceptive neurons. We were able to identify the receptive field for only five of the 10 C neurons. These five were clearly identifiable as mechanical nociceptive neurons, and intrasomal potential recording was successful (Fig. 1), except in one that was lost before measurement. Neurons responded with one spike to a pin prick and tended to become sensitized after repeated stimulation. There was no background activity. Average resting membrane potential (E_m) was −62.5 mV (± 6.0 S.D., $n = 4$), and AP amplitude (1) was 88 mV (± 10.9 S.D., $n = 4$). The average AP overshoot beyond the 0-mV level (2) was 25.8 mV (± 4.9 S.D., $n = 4$), AP duration at the E_m level (3) was 6.2 ms (± 1.7 S.D., $n = 4$), time to peak (4) was 2.7 ms (± 0.8 S.D., $n = 4$), peak rate of depolarization (5) was 67.5 V/s (± 14.8 S.D., $n = 4$), peak rate of repolarization (6) was 42.5 V/s (± 13.0 S.D., $n = 4$), after-hyperpolarization (AHP) height (7) was 14 mV (± 3.7 S.D., $n = 4$), and AHP duration to half-decay (8) was 15.0 ms (± 4.2 S.D., $n = 4$). The average CV was 0.95 m/s (± 0.4 S.D., $n = 5$).

The neurons responded to a pin prick (≥ 5 g) on the receptive field, which was almost round with a diameter of 1.5–2.0 mm. There was only one receptive field per neuron, and it was not sensitive to thermal stimulation. Somata from two neurons were successfully labeled with HRP (Fig. 2). The somata were small (22 μm × 20 μm, 20 μm × 18 μm) and almost round (Fig. 2A): the ratios of minor diameter to major were 0.91 and 0.90. Two neurons had no branching of the peripheral (p) axon in the ganglion. The set of three axons was unmyelinated: stem axons were 1.2, 1.0 μm in diameter, peripheral axons were 1.2, 1.0 μm in diameter, and central axons were 1.0, 0.8 μm in diameter. No branching in the ganglion was observed in peripheral (Fig. 2C) or central (Fig. 2D) axons.

The labeling of the central axon in the medulla oblongata stopped abruptly at the level of the nucleus interpolaris (TTDI) for one neuron, and it faded out in the root for another. Total lengths of the labeled axons were 1260 and 1480 μm to the edge of the section for peripheral axons and 1810 and 770 μm for central axons. No bifurcation of the central axon to the nucleus sensorius pricipalis n. trigemini (TPR) and no collaterals to the nucleus oralis (TTDO) or to the rostral half of the TTDI were

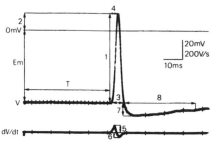

Fig. 1. Action potential of a C nociceptive neuron elicited by electrical stimulation demonstrating the wave-form parameters measured. Upper trace: membrane potential (V). Lower trace: time-differentiated response (dV/dt). T, latency from electrical stimulus = 46 ms (0.36 m/s); E_m, resting membrane potential (−78 mV). The following average values are listed in the text: AP amplitude (1), AP overshoot beyond the 0 mV level (2), AP duration at the E_m level (3), time to peak (4), peak rate of depolarization (5), peak rate of repolarization (6), AHP height (7), and AHP duration to half-decay to the E_m (8).

observed: this is very similar to a non-bifurcating fiber (central axon) of SC type of the A-delta nociceptive neuron [17].

We mainly compared the present data with those of our previous paper [17], because few other reports were suitable for comparison, lacking similar factors. According to these data the myelinated nociceptors were subdivided into two types: SC and FC. These are different from C nociceptive neurons (1) in CV, (2) in AP parameters, (3) in morphology of the soma, and (4) in having a set of three unmyelinated axons (stem, central and peripheral axons).

The mean CV of C nociceptors was less than 1 m/s and the fibers were unmyelinated, as reported by others (cf. [31]). The receptive fields of the present data were as small as those of C mechanical nociceptive neurons reported in earlier reports [3,6].

The E_m of C mechanical nociceptive neurons was similar to that of A-delta mechanical nociceptors. Although the spike height of C nociceptive neurons was similar to that of other A-delta nociceptors, the spike width was much wider. The peak rate of depolarization was especially low, and that of the AHP height was similar to that of other nociceptors.

The morphology of somata and axons of C ganglion neurons was investigated by several groups of researchers by intrasomal dye injection [11,13,16,30], but the present study is the first time that the small C neuron was proved to be a nociceptive neuron. It is doubtful that our C nociceptive neurons are the only kind of C neurons in the ganglion. Mouse A and H neurons, which have a small soma and unmyelinated axons, cannot be distinguished morphologically from each other [30]. Thus, it

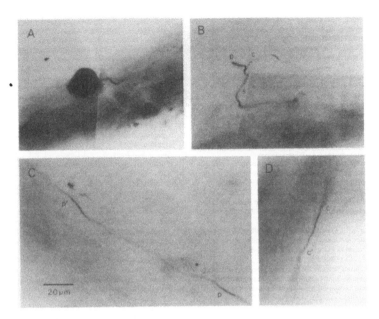

Fig. 2. Soma and axons of an HRP-labeled C nociceptive neuron. A, soma and stem axon; B, bifurcation of stem axon; C, peripheral axon; D, central axon. Abbreviations: c and c'(central side), central axon; p and p'(peripheral side), peripheral axon; s, stem axon. 80-μm horizontal section. The calibration bar in C is for all.

is possible that different types of C neurons exist in the snake trigeminal ganglion.

These somata of C nociceptive neurons were round, although other nociceptors in the A-delta class are elongated (0.76 for SC, 0.77 for FC, calculated from [17]). The mean major diameter of the soma of C mechanical nociceptive neurons was less than half that of A-delta nociceptive neurons. These C neurons were cells found in a group of the smallest in the ganglion, and were smaller than infrared (warm temperature) neurons ([14], *Agkistrodon*; [26], *Trimeresurus*).

The set of three axons of the C nociceptive neurons were all unmyelinated fibers of similar diameter. The difference in diameter between central and peripheral unmyelinated axons was reported as 50% by Suh et al. [25] and 55% by Lee et al. [16]. Although they did not define the modality, our data were inconsistent with theirs. Our results were also different from earlier studies stating that neurons with unmyelinated fibers have a central axon thinner than the peripheral and stem axons [5,10,13,22].

Our present data also contrast with the FC type of A-delta nociceptive neuron, which has a set of three myelinated axons with a thinner central axon, and with the SC type of A-delta nociceptive neuron, in which the central axon is unmyelinated and thinner than the other two.

In conclusion, a part of the C neuron group in the trigeminal ganglion was proved to be nociceptive neurons, which can be distinguished from the other two A-delta nociceptive neuron groups by physiological parameters (CV and AP parameters) and morphological characteristics (soma size and axon diameters).

We thank Dr. R.C. Goris for editing the English.

[1] Adams, J.C., Heavy metal intensification of DAB-based HRP reaction product, J. Histochem., 29 (1981) 775.

[2] Bessou, P., Burgess, P.R., Parl, E.R. and Taylor, C.B., Dynamic properties of mechanoreceptors with unmyelinated (C) fibers, J. Neurophysiol., 34 (1971) 116–131.

[3] Bessou, P. and Perl, E.R., Response of cutaneous sensory units with unmyelinated fibers to noxious stimuli, J. Neurophysiol., 32 (1969) 1025–1043.

[4] Cameron, A.A., Leah, J.D. and Snow, P.J., The electrophysiological and morphological characteristics of feline dorsal root ganglion cells, Brain Res., 362 (1986) 1–6.

[5] Gasser, H.S., Properties of dorsal root unmedullated fibers on the two sides of the ganglion, J. gen. Physiol., 38 (1955) 709–728.

[6] Georgopoulos, A.P., Functional properties of primary afferent units probably related to pain mechanisms in primate glabrous skin, J. Neurophysiol., 39 (1976) 71–83.

[7] Goris, R.C. and Nomoto, M., Infrared reception in oriental crotaline snakes, Comp. Biochem. Physiol., 23 (1967) 879–892.

[8] Görke, K. and Pierau, Fr.-K., Spike potentials and membrane properties of dorsal root ganglion cells in pigeons, Pflügers Arch., 386 (1980) 21–28.

[9] Gurtu, S. and Smith, P.A., Electrophysiological characteristics of hamster dorsal root ganglion cells and their response to axotomy, J. Neurophysiol., 59 (1988) 408–423.

[10] Ha, H., Axonal bifurcation in the dorsal root ganglion of the cat: a light and electron microscopic study, J. Comp. Neurol., 140 (1970) 227–240.

[11] Harper, A.A. and Lawson, S.N., Conduction velocity is related to morphological cell type in rat dorsal root ganglion neurons, J. Physiol., 359 (1985) 31–46.

[12] Harper, A.A. and Lawson, S.N., Electrical properties of rat dorsal root ganglion neurones with different peripheral nerve conduction velocities, J. Physiol., 359 (1985) 47–63.

[13] Hoheisel, U. and Mense, S., Observations on the morphology of axons and somata of slowly conducting dorsal root ganglion cells in the cat, Brain Res., 423 (1987) 269–278.

[14] Kishida, R., Terashima, S., Goris, R.C. and Kusunoki, T., Infrared sensory neurons in the trigeminal ganglia of crotaline snakes: transganglionic HRP transport, Brain Res., 241 (1982) 3–10.

[15] Koerber, H.R., Druzinsky, R.E. and Mendell, L.M., Properties of somata of spinal dorsal root ganglion cells differ according to peripheral receptor innervated, J. Neurophysiol., 60 (1988) 1584–1596.

[16] Lee, K.H., Chung, K., Chung, J.M. and Coggeshall, R.E., Correlation of cell body size, axon size, and signal conduction velocity for individually labelled dorsal root ganglion cells in the cat, J. Comp. Neurol., 243 (1986) 335–346.

[17] Liang, Y.-F. and Terashima, S., Physiological properties and morphological characteristics of cutaneous and mucosal mechanical nociceptive neurons with A-delta peripheral axons in the trigeminal ganglia of crotaline snakes, J. Comp. Neurol., 328 (1993) 88–102.

[18] Lieberman, A.R., Sensory ganglia, In D.N. Landon (Ed.), The Peripheral Nerve, Wiley, New York, 1976, pp. 188–278.

[19] Mei, N., Sensory structures in the viscera, Progr. Sens. Physiol., 4 (1983) 1–42.

[20] Molenaar, G.J., The sensory trigeminal system of a snake in the possession of infrared receptors. I. The sensory trigeminal nuclei, J. Comp. Neurol., 179 (1978) 123–136.

[21] Molenaar. G.J., The sensory trigeminal system of a snake in the possession of infrared receptors. II. The central projections of the trigeminal nerve, J. Comp. Neurol., 179 (1978) 137–175.

[22] Ranson, S.W., The structure of the spinal ganglia and of the spinal nerves, J. Comp. Neurol., 22 (1912) 159–169.

[23] Rose, R.D., Koerber, H.R., Sedivec, M.J. and Mendell, L.M., Somal action potential duration differs in identified primary afferents, Neurosci. Lett., 63 (1986) 259–264.

[24] Stoney Jr., S.D., Unequal branch point filtering action in different types of dorsal root ganglion neurons of frogs, Neurosci. Lett., 59 (1985) 15–20.

[25] Suh, Y.S., Chung, K. and Coggeshall, R.E., A study of axonal diameters and areas in lumbosacral roots and nerves in the rat, J. Comp. Neurol., 222 (1984) 473–481.

[26] Terashima, S. and Liang, Y.-F., Temperature neurons in the crotaline trigeminal ganglia, J. Neurophysiol., 66 (1991) 623–634.

[27] Terashima, S. and Liang, Y.-F., Physiological and morphological characteristics of C nociceptive neurons in the crotaline trigeminal ganglion, XXXII Int. Union Physiol. Sci. Abstr., 1993, pp. 93.

[28] Terashima, S. and Liang, Y.-F., Modality difference in the physiological properties and morphological characteristics of the trigeminal sensory neurons, Jpn. J. Physiol., Suppl. 1, 43 (1993) S267–S274.

[29] Traub, R.J. and Mendell, L.M., The spinal projection of individual identified A-delta- and C-fibers, J. Neurophysiol., 59 (1988) 41–55.

[30] Yoshida, S. and Matsuda, Y., Studies on sensory neurons of the mouse with intracellular-recording and horseradish peroxidase-injection techniques, J. Neurophysiol., 42 (1979) 1134–1145.

[31] Willis, W.D. and Coggeshall, R.E., Sensory Mechanisms of the spinal cord, Plenum, New York, 1979, 485 pp.

Touch and Vibrotactile Neurons in a Crotaline Snake's Trigeminal Ganglia

Shin-ichi Terashima[1] and Yun-Fei Liang

Department of Physiology, University of the Ryukyus School of Medicine, Nishihara-cho, Okinawa 903-01, Japan

Abstract   Thirty-five touch (M) neurons and 59 vibrotactile (V + M) neurons were recorded intrasomally in the trigeminal ganglion of a crotaline snake (the pit viper, *Trimeresurus flavoviridis*). The M neurons were excited by von Frey hair (5–10 mg) mechanical stimulation of the receptive field, and adapted slowly to a sustained stimulus. It was almost impossible to elicit 1:1 entrainment to sinusoidal movement. Vibration with touch was an adequate stimulus for the V + M neurons. The range of entrainment to sinusoidal movement was 5–300 Hz. Thresholds of V + M neurons to sustained mechanical stimulation could not be determined, but a response was obtained by stroking with a von Frey hair (5–10 mg). Receptive fields of both M and V + M neurons were found on the skin (scales) and the mucous membrane of the orofacial region. There was one receptive field of ~2 mm in diameter for each M or V + M neuron.

The mean resting potentials (± *SD*) of M and V + M neurons were −57.0 ± 5.1 mV (*n* = 26) and −63.7 ± 8.2 mV (*n* = 49), respectively. Neurons of both modalities displayed no background discharge. The action potential of V + M neurons had a shorter mean duration than that of M neurons. The mean conduction velocities (± *SD*) of peripheral (and stem) axons of M and V + M neurons were 28.4 ± 5.7 m/sec (*n* = 11) and 30.8 ± 7.8 m/sec (*n* = 30), respectively.

Recorded neurons were labeled with intrasomal horseradish peroxidase electrophoresis. V + M neurons had larger somata than M neurons. All axons of M and V + M neurons were myelinated and similar in diameter. M and V + M neurons had similar central projection patterns. The projection of the thick central axon divided into a thinner ascending fiber and a thick descending fiber at the entry zone of the root to the brainstem. The former ran ipsilaterally to the principal sensory nucleus of the trigeminal nerve (TPR), and the latter ran to the descending nucleus of the trigeminal nerve (TTD) and beyond, where terminal arbors and bouton swellings were observed. Smaller myelinated and unmyelinated collaterals were given off at right angles from the descending fiber of the central axon into the TTD. They projected more densely to the rostral part than to the caudal part of the TTD.

All of these data were compared with data on warm-temperature neurons, previously obtained.

Key words   trigeminal ganglion, somal action potential, touch neuron, axon diameter, central projection, mechanoreceptors, reptilian, vibrotactile sense organs

Somatosensory receptors have traditionally been classified according to their stimulus modality (mechanoreceptors, thermoreceptors, nociceptors, etc.). Since Ito's (1957, 1959) pioneering work on intrasomal recording from spinal ganglion neurons in the toad, many researchers have taken interest in the shape of spike potentials; the form of the action potential (AP) appears to be specific to the sensory

modality (lamprey: Martin and Wickelgren, 1971; Matthews and Wickelgren, 1978; Christenson et al., 1988; pigeon: Görke and Pierau, 1980; rat: Harper and Lawson, 1985b; cat: Cameron et al., 1986; Rose et al., 1986; Koerber et al., 1988), and to reflect the activity of ionic channels in the somal membrane (Matsuda et al., 1978, and Yoshida et al., 1978, in the mouse). More recently, Stoney (1990) categorized frog ganglion neurons on the basis of somatic spike shape, but these properties alone were not adequate to

---

1. To whom all correspondence should be addressed.

distinguish one type of sensory neuron from another without determining the adequate natural stimulus.

In morphological studies, two types of bifurcation of unipolar neurons have been reported in the ganglion (dog: Ranson, 1912; cat: Ranson and Davenport, 1931; Ha, 1970). Ito and Takahashi (1960) examined the relationship between the lengths of myelinated segments and the diameters of the three axons in the toad. Using intrasomal recordings, investigators have also reported a variety of morphological characteristics of the soma and its related axons (rat: Harper and Lawson, 1985a; cat: Lee et al., 1986; Hoheisel and Mense, 1987).

To trace central projections of a somatosensory neuron together with identifying the modality, intra-axonal recordings have been made from trigeminal axons in the medulla, followed by injection of horseradish peroxidase (HRP): nucleus principalis and nucleus oralis (Tsuru et al., 1989, and Shigenaga et al., 1990, in the cat), nucleus principalis to spinal dorsal horn C_1 and C_2 (Hayashi, 1985, in the rat), nucleus interpolaris (Jacquin et al., 1986, in the rat), and nucleus caudalis and spinal dorsal horn C_1 and C_2 (Jacquin et al., 1986, and Renehan et al., 1986, in the rat). All neurons of these projections were related to mechanoreception; however, on a single-neuron basis, combined morphological data and physiological data are still fragmentary, and an overall concept of the structural organization of mechanosensitive neurons is lacking.

In the present study, in order to find criteria for reliable modality identification of a neuron without determining the sensory response, we tried to obtain more systematic data on single touch (M) and vibrotactile (V + M) neurons of the trigeminal ganglion in the pit viper, *Trimeresurus flavoviridis*. We also compared the results with previously obtained data on warm-temperature neurons in this snake (Terashima and Liang, 1991). Once the modality identification is successful, data obtained from *in vitro* experiments and morphological investigation in which the soma is separated from its receptor may be related to modality. Thus, this knowledge may increase the efficiency of somatosensory research. Portions of this study have been published briefly elsewhere (Terashima et al., 1989; Terashima and Liang, 1993).

MATERIALS AND METHODS

Our preparation method and experimental protocol have been described in a previous paper (Terashima and Liang, 1991). Forty-nine pit vipers (31 males and 18 females, 250–450 g) were used in the present research. Anesthesia was induced with halothane. Pancuronium (2 mg/kg, i.m.) was administered for immobilization with artificial respiration when intrasomal recordings were made. Room temperature was kept at 24–26°C.

Each snake's skull was fixed with a head holder, and a hole was drilled through the inner ear to the trigeminal ganglia. An HRP-filled micropipette, as an intrasomal electrode, was inserted into the ganglion for recording potentials and for HRP injection by electrophoresis. The electrodes were filled with 4–6% HRP (Sigma, type VI) in 0.5 M KCl and 0.05 M Tris buffer (pH 7.3), and had a resistance of 60–100 MΩ at 1 kHz. The recorded potentials were amplified by a conventional electronic system and stored in a data recorder. APs were elicited by electric stimulation of the maxillary or mandibular nerve, and were recorded in the soma. The latency from the stimulus was measured to calculate the conduction velocity (CV) of APs in the peripheral axons. We were careful to find the minimum latency in the more favorable polarity and to use a stimulus just above threshold; the measured latency remained quite stable on repeated tests. The CV was calculated by dividing the distance between stimulating and recording electrodes by the latency. The criterion of acceptance for potential parameter measurement was for a neuron to show a stable resting membrane potential (E_m) exceeding -40 mV.

With a series of von Frey hairs as mechanical stimuli, we first identified M neurons that responded to 5 or 10 mg of stimulation. Then we confirmed the persistence of discharges (less than 2 sec) of slowly adapting M neurons while steady pressure was applied to each receptive field (RF) with a round-tipped metal rod (2 mm in diameter) mounted on a micromanipulator. We excluded pressure (P) neurons that had mechanical thresholds exceeding 100 mg and that showed a tendency to adapt rapidly. We also excluded nociceptive neurons that responded to more than 5 g of stimulation (Liang and Terashima, 1993, in *Trimeresurus*). Nociceptive neurons showed sensitization to repeated stimulation. The fourth category, V + M neurons, was not easy to discriminate from P neurons, which also appeared to give a rapidly adapting response. But such neurons could be differentiated by the presence of an off-response at the cessation of a sustained mechanical stimulus. Although the responses were phasic, unlike those in nociceptive neurons (Liang and Terashima, 1993, in *Trimeresurus*), the off-response was characteristic of the V + M modality and served to distinguish these neurons from other neurons with mechanical sensitivity. V + M neurons responded to a sustained mechanical stimulus with a single discharge, and M neurons responded to such a stimulus with continuous discharge. The fifth category, vibration (V) neurons, had a large RF that was usually not round. They responded to mechanical stimulation, as the V + M neurons did, when stimulated at the most sensitive spot, which was usually the center of the RF; there was a sensitivity gradient in the RF. The boundary was not distinct, as in the V + M neurons. Data on P neurons and V neurons are not reported in the present paper because the sample number was insufficient for analysis.

A hand-held vibrator was made from a dynamic speaker with a glass stylus (2 mm in diameter) glued to the center. The tip of the stylus was 200 μm in diameter. Sinusoidal waves were supplied by a function generator (Iwatsu SG-4101), powered by an amplifier the output of which was fed to the vibrator. Sinusoidal movement of the stylus was monitored by the output of a strain gauge through a Wheatstone bridge. The peak-to-peak amplitude was 350 μm. The indentation in the skin was not measured. The strain gauge was attached between the stylus and the speaker frame. The distortion of the sinusoidal movement was negligible between 1 and 1000 Hz. The range used for stimulation was 1 to 500 Hz, because entrainment of a V + M neuron could not be obtained above 500 Hz. We scanned the range to find the best entrainment frequency (Proske, 1969b, in *Pseudechis*) for each neuron. Either forceps or a metal rod (2 mm in diameter) warmed in hot water (45–70°C) or chilled in ice water (0°C) was used for detecting temperature sensitivity (warm and cold). Vibrating forceps were used for detecting vibratory modality for the screening test. Jaw movement was tested during intrasomal recording (Terashima et al., 1989, in *Trimeresurus*).

After potentials were recorded, HRP was intrasomally injected by applying positive current; the current was alternately turned on for 1500 msec (10–30 nA) and off for 500 msec for 15–20 min. Only one ganglionic neuron on each side of the head was injected, in order to avoid confusion in tracing the axons.

After 24–48 hr of survival, each snake's head was perfused with a solution via the carotid arteries on both sides. The perfusate was first 0.75% saline with 7 IU heparin/ml, then a fixative (2% paraformaldehyde and 2.5% glutaraldehyde in 0.1% phosphate buffer [PB], pH 7.4, 10°C), 0.1 M PB (pH 7.4, 10°C), and finally 10% sucrose in PB for more than 20 min. The distance between the stimulation site on the nerve and the recording site on the ganglion was then measured to calculate the CV. The brainstem and ganglia were removed and stored in 30% sucrose in PB at 10°C overnight. Frozen horizontal sections were cut at 40 μm and reacted with diaminobenzidine according to a method modified by Adams (1981). The sections were mounted on chrome alum–gelatin slides, dried for 3 hr, and counterstained with cresyl violet; they were then observed with an Olympus BH-2 microscope. Histological identification and nomenclature follows that of Molenaar (1978a,b) for snakes. Photomicrographs were taken with an exposure control unit (Olympus PM-CBSP). A camera lucida was used for sketching neurons. The diameters of the axons and somata were measured with an ocular micrometer, which was calibrated with a stage micrometer. The diameters of the central, peripheral, and stem axons were measured across the outside of the fibers within the ganglion at 50-μm intervals from T-bifurcations. Ten measurements were made on peripheral axons and five on central and stem axons, and

the mean values were taken as their size. For the measurement of the diameter of a soma, the largest cross-sectional area was always chosen.

Probability values were tested by Cochran and Cox's (1957) two-tailed t test.

RESULTS

Mode of Response

We recorded intrasomal responses from 35 M and 59 V + M neurons in the second and third divisions of the trigeminal ganglion. None displayed background discharge at room temperature (25°C), and all responded to mechanical stimulation of the skin or mucosa and to electrical stimulation of the nerve. They did not respond to either heat or cold, and responded with the same pattern of discharge to both heated and cooled metal displacement.

The response of the M neurons continued at a frequency of up to 29.4 Hz during sustained mechanical stimulation (Fig. 1A), and adapted slowly for less than 2 sec, but we could not determine the upper frequency limit. For M neurons it was impossible to elicit a prolonged 1:1 response with sinusoidal movement at any frequency. Sustained mechanical stimulation elicited brief, regular firing.

V + M neurons, which were sensitive to vibration with touch, were identified in the trigeminal ganglia. These neurons responded well to vibratory stimulation when the tip of the vibrator rod touched the RF, but vibration applied outside the RF on the same scale of the snake had no effect. Because these were rapidly adapting receptors, persistent pressure generally produced no response, but it occasionally caused an off-response. A light tap on the frame on which the animal was mounted elicited no response.

For all 40 V + M neurons in which the frequency range for entrainment was examined, spike discharges synchronized 1:1 with the stimulus cycle from 5 to 300 Hz when the vibrating stylus was maintained in an optimum position (Fig. 1B). Outside of this range, neuronal response was erratic, with entrainment at 1:2, 1:3, or 1:4. Absolute thresholds and tuning curves for this frequency range were not determined. The compression phase of sinusoidal stimulation was more effective than the release phase.

Intrasomal Recording of Resting Membrane Potentials

The mean E_m of M neurons was −57.0 mV, and that of V + M neurons was −63.7 mV. These E_ms were significantly different from each other (Table 1).

Conduction Velocities of Peripheral Axons

Electrical stimulation of the superficial branch of the maxillary nerve produced one spike in the soma (Fig. 2). APs

A

B

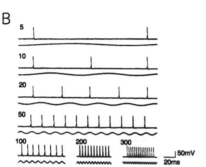

FIGURE 1. Examples of somatic APs of an M neuron (A) and a V + M neuron (B) evoked by natural stimulation applied to the RF. (A) The upper trace is the membrane potential; the E_m was −48 mV. The lower trace indicates a period of sustained touch stimulation of ~2 g. (B) The upper trace in each record represents the neuron discharge; the E_m was −64 mV. The lower trace is the stimulus marker of the applied sinusoidal vibration. The skin is indented during the upward phase. The frequency of vibration (Hz) is given in the upper left corner of each record.

were elicited by electrical stimulation and showed latencies of 0.7 msec and 0.6 msec at the recorded soma for M and V + M neurons, respectively (Fig. 2).

For M neurons, the mean CV (± SD) was 28.4 ± 5.7 m/sec (n = 11); for V + M neurons, it was 30.8 ± 7.8 m/sec (n = 30). Judging from our previous data on compound APs in *Trimeresurus* (Terashima and Liang, 1991), the values of M and V + M neurons placed them in the A-beta group, the fastest group in the cutaneous sensory nerves of this species.

Measurement of Action Potential Parameters

The AP of M and V + M neurons was shaped like the antidromic evoked spike of a motoneuron (cf. Eccles, 1957)—a thin spike with no hump on the falling phase. The mean AP parameters of M and V + M neurons were measured as shown in Figure 2 and are given in Table 1. Statistical analysis of M and V + M neuron responses indicated significant differences for AP amplitude, AP duration at E_m level, time to peak, peak rate of depolarization, and peak rate of repolarization. M afferents had significantly lower AP amplitude, longer AP duration at E_m level, longer time to peak, smaller peak rate of depolarization, and smaller peak rate of repolarization than V + M neurons. M neurons were statistically similar to V + M neurons in afterhyperpolarization (AHP) height and AHP duration to half-decay (Table 1).

TABLE 1. Physiological Properties of Touch and Vibrotactile Neurons

Measurements	M	V + M	M vs. V + M
AP amplitude (mV)	68.08 ± 8.29 (26)	77.8 ± 11.5 (59)	<0.001*
AP overshoot beyond the 0-mV level (mV)	11.07 ± 6.16 (26)	14.3 ± 7.4 (55)	>0.05
AP duration at E_m level (msec)	1.68 ± 0.23 (16)	1.3 ± 0.2 (29)	<0.001*
Time to peak (msec)	0.64 ± 0.10 (16)	0.5 ± 0.1 (29)	<0.001*
Peak rate of depolarization (V/sec)	209.0 ± 29.0 (13)	356.8 ± 83.3 (28)	<0.001*
Peak rate of repolarization (V/sec)	−102.0 ± 10.0 (13)	−175.7 ± 53.9 (28)	<0.001*
AHP height (mV)	−5.06 ± 1.52 (16)	−4.3 ± 2.80 (29)	>0.2
AHP duration to half-decay (msec)	1.75 ± 0.35 (16)	1.5 ± 0.4 (23)	>0.05
E_m (mV)	−57.0 ± 5.1 (26)	−63.7 ± 8.2 (49)	<0.001*

Note. Values are means ± SD. Numbers in parentheses indicate number of neurons. AP, action potential; E_m, resting membrane potential; AHP, afterhyperpolarization. A two-tailed t test was used.

* Significant difference between means.

A

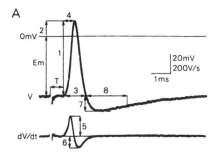

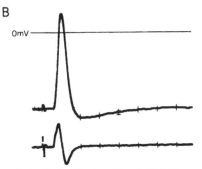

B

FIGURE 2. An AP of an M neuron (A) and a V + M neuron (B) elicited by electrical stimulation. Upper trace: Membrane potential (V). Lower trace: Time-differentiated response (dV/dt). (A) The waveform parameters measured: T, the latency, was 0.7 msec (22.9 m/sec), and E_m was -62 mV. The following average values are listed in Table 1: AP amplitude (1), AP overshoot beyond the 0-mV level (2), AP duration at the E_m level (3), time to peak (4), peak rate of depolarization (5), peak rate of repolarization (6), AHP height (7), and AHP duration to half-decay to the E_m (8). (B) Same scale as in A. Latency was 0.6 msec (27.5 m/sec), and E_m was -78 mV.

Receptive Field Properties

RFs of M neurons were located in the orofacial region on the skin or on the mucosa (Fig. 3). The average area was small (2.1 mm × 2.1 mm; $n = 35$). The RFs had clear borders and homogeneous sensitivity within the area; the mechanical thresholds were in the 5- to 10-mg range. There was only one RF per neuron, and the area was unchanged after electrophoresis.

Eighteen RFs of V + M neurons were delineated on the scales or on the oral mucosa (Fig. 3). The area sizes ranged from 1.5 mm × 1.5 mm to 3.5 mm × 3.5 mm ($n = 59$). There was usually only one area limited by the border of a scale, but sometimes it was as wide as two scales and at other

times it occupied only half of one scale. No visible difference was found on the skin surface of the RF, but the border of an RF on the scale was clearly detectable with vibrator stimulation. Although V + M neurons were insensitive to steady pressure, a response could be obtained by stroking with 5- to 10-mg von Frey hairs.

The RFs of M and V + M neurons were not restricted to any specific region (Fig. 3). We could not find cutaneous corpuscles like those in the Texas rat snake (Jackson and Doetsch, 1977a), or a dome RF such as an Iggo corpuscle in the cat (Iggo and Muir, 1969).

Morphological Characteristics of Somata

After intrasomal recording, we injected HRP by electrophoresis for cell labeling (Figs. 4 and 5). Nerve fibers ran through the center of the ganglion, and sensory cells surrounded them, causing a swelling. Somata of M and V + M neurons were 47.0 ± 12.6 μm ($n = 8$) and 52.6 ± 10.1 μm ($n = 14$) in major diameter, and 27.4 ± 6.6 μm ($n = 8$) and 37.4 ± 8.2 μm ($n = 14$) in minor diameter, respectively (Table 2). The difference in major diameter between the two was not statistically significant ($p > 0.2$), but that of the minor diameter was ($p < 0.01$): V + M somata were larger than M somata. The ratio of minor diameter to major diameter for M and V + M neurons was 0.62 ± 0.19 ($n = 8$) and 0.71 ± 0.15 ($n = 14$), respectively. The somata were ellipsoidal in shape, with their long axis directed toward the stem axon.

Morphological Characteristics of Axonal Bifurcations

Although the course of the stem axon in the ganglion was not as smooth as that of the peripheral and the central axons, there were no glomerular structures. Axons had a constriction at the bifurcation point (Figs. 4B, 5), but did not have a triangular expansion. Ranvier's nodes were observed at branch points. Although we could not see the myelin sheath, we interpreted these nodes as indications of myelination. The nodes were more numerous near the bifurcation (Fig. 5). All three branches (i.e., stem, peripheral, and central axons) of M and V + M neurons were large myelinated (A) fibers (Figs. 4B, 5; Table 2). There was no significant difference in diameter among the three branches of any given neuron. Additional bifurcation of the peripheral axon in the ganglion was not observed for M and V + M neurons. Tapering of the peripheral axon in the ganglion also was not observed, and it was not well labeled far from the ganglion.

Morphological Characteristics of Central Projections

Projections of six M and nine V + M neurons were fully traced to their central terminals (Fig. 6), and both types

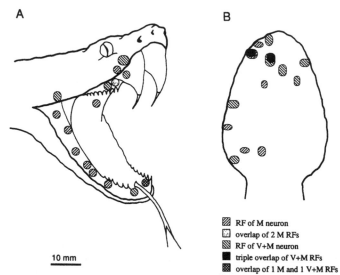

A

B

10 mm

RF of M neuron
overlap of 2 M RFs
RF of V+M neuron
triple overlap of V+M RFs
overlap of 1 M and 1 V+M RFs

FIGURE 3. Distribution of RFs of 17 M neurons and 18 V + M neurons. (A) Snake's head. (B) Ventral side of the jaw.

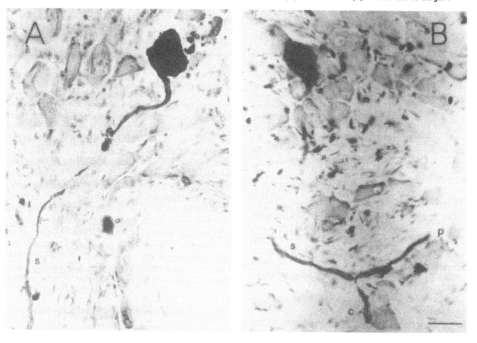

FIGURE 4. Photomicrographs of an HRP-labeled M neuron (A) and a V + M neuron (B) (40-μm horizontal sections). An arrow in B indicates bifurcation of stem axon: stem (s), central (c), and peripheral (p) axon. Calibration bar: 40 μm.

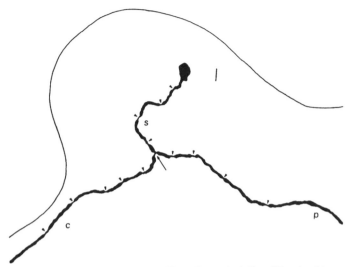

FIGURE 5. Camera lucida drawing of the V + M neuron shown in Figure 4B. An arrow indicates bifurcation of the stem axon. Arrowheads indicate nodes of Ranvier. The upper curved line indicates the ganglion margin. s, stem axon; p, peripheral axon; c, central axon. Calibration bar: 40 μm.

projected to the same central nuclei (Figs. 7, 8), although there was much variety in arborization in the nuclei.

We could not find any specific difference between M and V + M neurons in the shape of their terminal arbors or the frequency of collaterals. The central axons of M and V + M neurons bifurcated into a thin ascending fiber and a thick descending fiber at the entrance of the root to the brainstem (Figs. 7, 8). The diameter of ascending fibers of M and V + M neurons (means = 3.3 and 2.8 μm, respectively) was significantly smaller than that of the central axons (means = 6.0 and 5.5 μm, respectively) and descending fibers (means = 5.8 and 6.0 μm, respectively) (Table 3).

TABLE 2. Axon and Soma Sizes of Touch and Vibrotactile Neurons

Measurement	M	V + M	M vs. V + M
Stem axon (μm)	5.67 ± 1.89 (9)	5.0 ± 1.5 (16)	>0.2
Central axon (μm)	5.63 ± 2.06 (8)	4.6 ± 1.9 (15)	>0.2
Peripheral axon (μm)	5.78 ± 1.75 (9)	4.9 ± 1.6 (15)	>0.2
Major diameter (μm)	47.0 ± 12.6 (8)	52.6 ± 10.1 (14)	>0.2
Minor diameter (μm)	27.4 ± 6.6 (8)	37.4 ± 8.2 (14)	<0.01*
p values			
Stem vs. central	>0.8	>0.5	
Stem vs. peripheral	>0.8	>0.8	
Central vs. peripheral	>0.8	>0.5	

Note. Values are means ± SD. Numbers in parentheses indicate number of neurons. A two-tailed *t* test was used.

* Significant difference between means.

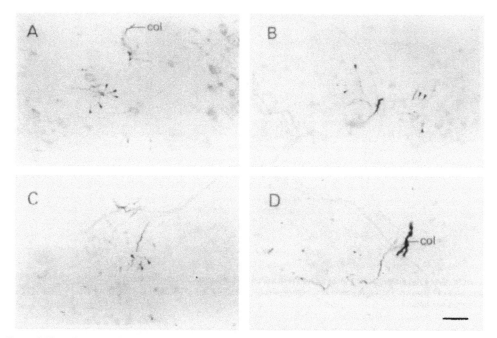

FIGURE 6. Photomicrographs of terminal arbors of an M neuron (A and B) and a V + M neuron (C and D) in the trigeminal nervous system (40-μm horizontal sections). (A) TPR (principalis). (B) TTDO (oralis). (C) TTDI (interpolaris). (D) TTDC (caudalis). Arrowheads indicate bouton swellings. col, collateral. Upper side is lateral, and left side is caudal. Calibration bar: 40 μm.

The ascending fiber projected only to the ipsilateral nucleus sensorius principalis n. trigemini (TPR). The descending fiber projected to the ipsilateral nucleus descendens n. trigemini (TTD) and beyond. This nucleus was composed of the nucleus oralis n. trigemini (TTDO), the nucleus interpolaris n. trigemini (TTDI), and the nucleus caudalis n. trigemini (TTDC) (Figs. 6, 7, 8). Smaller myelinated and unmyelinated collaterals branched off from the descending fiber at right angles.

Terminal arbors and terminal boutons were observed in the nuclei forming terminal foci, with some adjacent terminal fields overlapping. Terminals were most numerous in the TPR, and their numbers became successively smaller from the TTDO to the TTDC. The descending fiber tapered in its course, especially between the TTDI and the TTDC.

The final destination of the descending fiber of the central axon could not be confirmed in the present study. Terminal arbors in the TTDC were usually very fine and short, and could not be photographed optimally. Collaterals to Cajal's interstitial nucleus (Molenaar, 1978a,b, in the python), also called the paratrigeminal nucleus (Phelan and Falls, 1989, in the rat), were not observed.

DISCUSSION

Although our method of modality identification was based on gram-force thresholds, a comparison with other reptilian data seems worthwhile. Jackson and Doetsch (1977b) reported three classes of noncorpuscular mechanoreceptors in snakes—rapidly adapting, slowly adapting, and intermediate-adapting—in addition to a specialized corpuscle in the skin (1977a, in *Elaphe obsoleta*). All of their mechanical thresholds were more than 10 times higher than ours, although this might be attributable to methodological differences. They found no significant difference in threshold among the three classes, as between our M and V + M receptors.

The slowly adapting receptors in snakes reported by Proske (1969a, in *Pseudechis*), which had a low threshold to mechanical stimulation and were insensitive to vibration, seem to be equivalent to our M neurons, although they became sensitive to temperature change (much like a cold receptor) when they were stimulated mechanically.

Two classes of slowly adapting neurons in reptiles, type I and type II, similar to mammalian slowly adapting neurons, were reported by Kenton et al. (1971) in the alligator. These neurons seem to be functionally analogous to our M neurons. Although ours had no thermal sensitivity, their finding of cooling responses in slowly adapting mechanoreceptors was perhaps an artifact of the visually observable mechanical effect induced by ethyl chloride spray (L. Kruger, personal communication). Thus, the polymodal status of their neurons is questionable.

Rapidly adapting receptors in Proske's (1969a) report on *Pseudechis* seem to be equivalent to our V + M neurons, in that (1) they did not respond to extraneous stimuli, such as a light tap on the base plate supporting the experimental animal; (2) they responded up to 300 Hz; (3) they were best stimulated by scratching or scraping the scales; and (4) RFs were discrete and restricted to one or two scales. Siminoff

and Kruger's (1968) rapidly adapting fibers in the alligator also seem to be the same as our V + M neurons, because they were sensitive to small mechanical transients and responded to vibrations of relatively low frequency (100–200 Hz).

Electrophysiological Properties

Intrasomal recording of potentials in ganglion cells has been performed *in vivo* (Görke and Pierau, 1980, in the pigeon), and the modality of such recorded neurons has also been confirmed (Rose et al., 1986, and Koerber et al., 1988, in the cat). Along with the intrasomal recording, intrasomal labeling has also been done both *in vitro* (cat: Lee et al., 1986; rat: Harper and Lawson, 1985a; mouse: Yoshida and

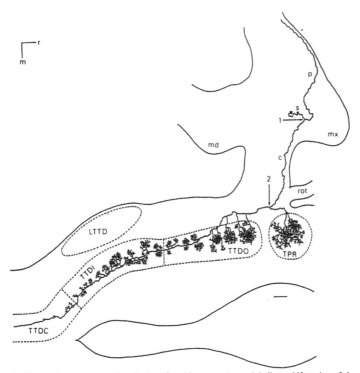

FIGURE 7. Camera lucida drawing of the central projection of an M neuron. Arrow 1 indicates bifurcation of the stem axon in the trigeminal ganglion, and arrow 2 indicates bifurcation of the central axon at the entrance to the brainstem. LTTD, nucleus descendens lateralis n. trigemini; TPR, nucleus sensorius principalis n. trigemini; TTDC, nucleus caudalis n. trigemini; TTDI, nucleus interpolaris n. trigemini; TTDO, nucleus oralis n. trigemini; c, central axon; m, medial; md, mandibular branch of the trigeminal nerve; mx, maxillary branch of the trigeminal nerve; p, peripheral axon; r, rostral; rot, radix ophthalmicus n. trigemini; s, stem axon. Calibration bar: 200 μm.

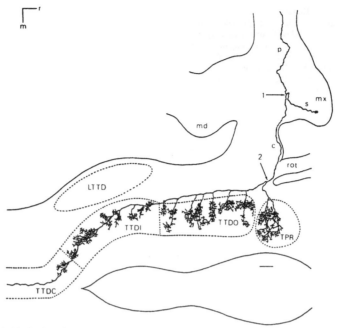

FIGURE 8. Camera lucida drawing of central projection of a V + M neuron. Arrow 1 indicates bifurcation of the stem axon in the trigeminal ganglion, and arrow 2 indicates bifurcation of the central axon at the entrance to the brainstem. Abbreviations and calibration bar as in Figure 7.

Matsuda, 1979) and *in vivo* (Cameron et al., 1986, and Hoheisel and Mense, 1987, in the cat). Our current work (Terashima and Liang, 1991, and Liang and Terashima, 1993, in *Trimeresurus*) is the first to report successful intrasomal recording together with labeling of a sensory ganglion neuron from the soma to its terminals in the reptilian central nervous system (CNS).

For warm-temperature (T) neurons and neurons responding to both warm temperature and mechanical stimulation (T + M neurons), eight physiological parameters have

TABLE 3. Fiber Diameters of the Second Bifurcation at the Pons

Measurement	M	V + M	M vs. V + M
Central axon (μm)	6.0 ± 2.16	5.5 ± 1.50	>0.5
	(6)	(6)	
Ascending fiber (μm)	3.3 ± 1.75	2.8 ± 1.65	>0.5
	(6)	(6)	
Descending fiber (μm)	5.8 ± 1.95	6.0 ± 2.24	>0.8
	(6)	(6)	
p values			
Central vs. ascending	<0.05*	<0.05*	
Central vs. descending	>0.8	>0.5	
Ascending vs. descending	<0.05*	<0.05*	

Note. values are meands ± *SD*. Numbers in parentheses indicate number of neurons. A two-tailed *t* test was used.

* Significant difference between means.

already been shown to be similar (Terashima and Liang, 1991). Therefore, we compared this T group (both T and T + M neurons) with the M and V + M neurons. Six AP parameters out of eight (AP amplitude, AP overshoot beyond the 0-mV level, AP duration at E_m level, time to peak, peak rate of depolarization, and peak rate of repolarization) were significantly different, except for the AP amplitude and AP overshoot between M and T, and the AP overshoot between M and T + M. Generally V + M spikes were the highest and narrowest, and T-group spikes the lowest and broadest; M neurons tended to fall between V + M and T-group neurons with respect to AP duration at E_m level, time to peak, peak rate of depolarization, and peak rate of repolarization.

The CVs of M-group neurons (M and V + M) were faster than those of T-group neurons. This might be related to the axon diameter. The CVs in the stem axon of tactile neurons were measured at 0.5–2 m/sec by Darian-Smith et al. (1965) in the cat, and those of the peripheral axon ranged from 11.9 to 61.5 m/sec. If the CV of an impulse is proportional to the axon diameter and their tactile neurons had three branches of similar diameter, then the CVs of the stem and peripheral axons should have been very similar.

There were significant differences between V + M and M neurons in E_m and in all the AP parameters except AP overshoot and the AHP (Table 1). The peak rate of rise (dV/dt) was significantly larger in M-group neurons than in T-group neurons, but AHP duration and height were smaller. The combination of large dV/dt and small AHP is consistent with the report of Koerber et al. (1988), who compared high- and low-threshold mechanical receptors in the cat.

Yoshida et al. (1978) recorded intrasomal potentials *in vitro* in the mouse, and classified the neurons by spike form into three kinds (F, H, and A), the forms of which were related to the ionic channels of the soma membrane. None of their potentials were as high as in our present results, but judging from their narrow spike width and the absence of a hump on the falling phase of the AP, our M and V + M neurons correspond to their F neurons.

Receptive Field Properties

The RFs of noncorpusclar mechanoreceptors in Texas rat snakes (Jackson and Doetsch, 1977b) were larger than those of M and V + M receptors, and usually lower stimulus thresholds were located near the center, which suggests a possibility of correspondence with our V neurons (see "Materials and Methods").

Morphological Characterization

We compared the morphological characteristics of somata and axons in M, V + M, and T-group neurons. In previous work, the somata of T and T + M neurons were shown to be similar in size (Terashima and Liang, 1991). When M-

group neurons were compared with T-group neurons, only the major diameter of V + M somata was significantly different from that of T-group neurons. V + M neurons were the largest neurons in this ganglion.

There are large light and small dark neurons in the spinal ganglion (Hatai, 1902, in the rat; Lawson, 1979, in the mouse). These differences are also found in the snake trigeminal ganglion (Kishida et al., 1982, in *Agkistrodon*). The large neurons are of placodal origin, and the small ones are of neural crest origin (Hamburger, 1961, in the chick embryo). In the present experiment V + M neurons had the largest cell bodies, so they are possibly of placodal origin. It is also possible that a thick central axon, as in M and V + M neurons, is a morphological characteristic of placodal origin.

It has been reported that large cells have large peripheral axons, and vice versa (Cameron et al., 1986, in the cat; Yoshida and Matsuda, 1979, in the mouse). Ours did not, because the somata of M neurons were not significantly different from that of T-group neurons, whereas the peripheral axons of M-group neurons were significantly larger. Poor correlation of CV with cell soma size in fast-conducting A neurons was reported by Harper and Lawson (1985a) in the rat. Hoheisel and Mense (1987) reported a great overlap among somata within A-neuron groups in the cat. Our results were consistent with these findings. The soma size of a sensory neuron is not suitable for distinguishing its modality.

The finding that M and V + M neurons had three branches of similar diameter (Table 2) is consistent with other observations (Ha, 1970, in the cat; Ranson, 1912, in the dog), but this is the first time that bifurcation of the M and V + M neurons has been shown to be of this type. This similarity in branch diameter is characteristic of M-group neurons, and significantly different from that of T-group neurons; the difference in central axon diameter between the M group and the T group seems to be modality-specific.

In their work on toads, Ito and Takahashi (1960) reported a low safety factor in centrifugal conduction to the bifurcation, and attributed it to the small diameter of the central axon. Lee et al. (1986) reported that the central axon of A neurons in the cat, like that of C neurons, has a smaller diameter than the peripheral axons. This is inconsistent with the findings of our present study on A-beta neurons (M and V + M), but consistent with our data on A-delta neurons (T and T + M). The morphology of the related axons has been proved to be modality-specific in the T-group neurons in *Trimeresurus* (Terashima and Liang, 1991). This might provide a set of important correlates for distinguishing the stimulus modality of a sensory neuron.

Central Projections

Windle (1926) observed three kinds of trigeminal fibers in the medulla of the mouse and fetal pig, and suggested that a touch neuron sends information to the TPR by way of ascending nonbifurcating fibers or to the TPR and TTD by way of bifurcating fibers. In the present study, M and V + M neurons had bifurcating fibers but did not have ascending nonbifurcating fibers. Our observation that descending axons taper in the tractus descendens n. trigemini (ttd) has also been suggested by Darian-Smith et al. (1965) from the observation that the CV slows in the medulla of the cat.

Projection patterns of M and V + M fibers in the medulla oblongata were similar, but were quite different from those of T-group neurons, which project only to the nucleus descendens lateralis n. trigemini (LTTD) (Terashima and Liang, 1991). Unlike M and V + M afferents, T-group afferents gave rise to small collaterals with small terminal arbors characterized by the absence of right-angle branching. The density of T-group neurons' arborization in the nucleus lies between that of M-group neurons in the TTDI above and in the TTDC below. The similarity in sites of central projection between M and V + M neurons is consistent with Hayashi's (1985) observation concerning mechanoreceptors: that large myelinated axons, which include hair and slowly adapting mechanical receptors, project to the equivalent nuclei in the rat. The absence of nonbifurcation of the central branch of our M-group neurons at the root entry zone may reflect the small sample size.

M and V + M neurons seem to share information processing in the CNS, because they share the same nuclei at the projection site. However, they may constitute a dual sensory system in the sense of being served by two distinct sets of primary afferent neurons: one most sensitive to frequencies of 5–40 Hz, and a second sensitive over the range of 60–300 Hz (Talbot et al., 1968, in the human and monkey). From this point of view, V + M and M neurons might be called mechanoreceptive submodalities.

When we compared central projections of M and V + M neurons with those of T-group neurons, not only electrophysiological properties, but also the thickness of the central axon and the projection nuclei, were modality-specific. Although T + M neurons are mechanically sensitive, with a threshold as low as that of M neurons (Terashima and Liang, 1991, 1993), T-group neurons of both submodalities project only to the LTTD. When we consider these results, the central projection pattern seems to present a more basic relationship of sensory mechanisms than can be deduced simply from the mode of response in bimodal neurons.

In conclusion, the morphological characteristics of M and V + M neurons were similar, although some physiological characteristics were significantly different. Judging from the similarity of the central projections, there may be similar information processing in the CNS for M and V + M modalities. With regard to identifying the submodality specificity of an individual somatosensory neuron in the absence of knowledge of the adequate stimulus, we conclude that the following characteristics must be taken into account:

(1) spike form; (2) axon diameter of the set of three branches, especially thickness of the central axon; (3) central projection pattern, and (4) CV. With knowledge of these four characteristics, we can reliably predict a neuron's submodality sensitivity without observing its response to natural stimulus. Soma size is not suitable for identification of somatosensory modality because of extensive modality overlap.

ACKNOWLEDGMENTS

We thank Dr. L. Kruger for critical reading and comments on the manuscript, Dr. R. C. Goris for editing the English, and Ms. Ai-Qing Zhu for assisting with our experiments.

REFERENCES

ADAMS, J. C. (1981) Heavy metal intensification of DAB-based HRP reaction product. J. Histochem. Cytochem. 29: 775.

CAMERON, A. A., J. D. LEATH, and P. J. SNOW (1986) The electrophysiological and morphological characteristics of feline dorsal root ganglion cells. Brain Res. 362: 1–6.

CHRISTENSON, J., A. BORMAN, P.-Å. LANGERBÄCK, and S. GRILLNER (1988) The dorsal cell, one class of primary sensory neuron in the lamprey spinal cord: I. Touch, pressure but no nociceptior—a physiological study. Brain Res. 440: 1–8.

COCHRAN, W. G., and G. M. COX (1957) Experimental Designs, Wiley, New York.

DARIAN-SMITH, I., P. MUTTON, and R. PROCTOR (1965) Functional organization of tactile cutaneous afferents within the semilunar ganglion and trigeminal spinal tract of the cat. J. Neurophysiol. 28: 682–694.

ECCLES, J. C. (1957) The Physiology of Nerve Cells, Oxford University Press, London.

GÖRKE, K., and F.-K. PIERAU (1980) Spike potentials and membrane properties of dorsal root ganglion cells in pigeons. Pflügers Arch. 386: 21–28.

HA, H. (1970) Axonal bifurcation in the dorsal root ganglion of the cat: A light and electron microscopic study. J. Comp. Neurol. 140: 227–240.

HAMBURGER, V. (1961) Experimental analysis of the dual origin of the trigeminal ganglion in the chick embryo. J. Exp. Zool. 148: 91–124.

HARPER, A. A., and S. N. LAWSON (1985a) Conduction velocity is related to morphological cell type in rat dorsal root ganglion neurones. J. Physiol. (Lond.) 359: 31–46.

HARPER, A. A., and S. N. LAWSON (1985b) Electrical properties of rat dorsal root ganglion neurones with different peripheral nerve conduction velocities. J. Physiol. (Lond.) 359: 47–63.

HATAI, S. (1902) Number and size of the spinal ganglion cells and dorsal root fibers in the white rat at different ages. J. Comp. Neurol. 12: 107–124.

HAYASHI, H. (1985) Morphology of central terminations of intra-axonally stained, large, myelinated primary afferent fiber from facial skin in the rat. J. Comp. Neurol. 237: 195–215.

HOHEISEL, U., and S. MENSE (1987) Observations on the morphology of axons and somata of slowly conducting dorsal root ganglion cells in the cat. Brain Res. 423: 269–278.

IGGO, A., and A. R. MUIR (1969) The structure and function of a slowly adapting touch corpuscle in hairy skin. J. Physiol. (Lond.) 200: 763–796.

ITO, M. (1957) The electrical activity of spinal ganglion cells investigated with intracellular microelectrodes. Japan. J. Physiol. 7: 297–323.

ITO, M. (1959) An analysis of potentials recorded intracellularly from the spinal ganglion cell. Japan. J. Physiol. 9: 20–32.

ITO, M., and I. TAKAHASHI (1960) Impulse conduction through spinal ganglion. In *Electrical Activity of Single Cells*, Y. Katsuki, ed., pp. 159–179, Igaku Shoin, Tokyo.

JACQUIN, M. F., W. E. RENEHAN, R. D. MOONEY, and R. W. RHOADES (1986) Structure–function relationships in rat medullary and cervical dorsal horns: I. Trigeminal primary afferents. J. Neurophysiol. *55*: 1153–1186.

JACKSON, M. K., and G. S. DOETSCH (1977a) Functional properties of nerve fibers innervating cutaneous corpuscles within cephalic skin of the Texas rat snake. Exp. Neurol. *56*: 63–77.

JACKSON, M. K., and G. S. DOETSCH (1977b) Response properties of mechanosensitive nerve fibers innervating cephalic skin of the Texas rat snake. Exp. Neurol. *56*: 78–90.

KENTON, B., L. KRUGER, and M. WOO (1971) Two classes of slowly adapting mechanoreceptor fibers in reptile cutaneous nerve. J. Physiol. (Lond.) *212*: 21–44.

KISHIDA, R., S. TERASHIMA, R. C. GORIS, and T. KUSUNOKI (1982) Infrared sensory neurons in the trigeminal ganglia of crotaline snakes: Transganglionic HRP transport. Brain Res. *241*: 3–10.

KOERBER, H. R., R. E. DRUZINSKY, and L. M. MENDELL (1988) Properties of somata of spinal dorsal root ganglion cells differ according to peripheral receptor innervated. J. Neurophysiol. *60*: 1584–1596.

LAWSON, S. N. (1979) The postnatal development of large light and small dark neurons in mouse dorsal root ganglia: A statistical analysis of cell numbers and size. J. Neurocytol. *8*: 275–294.

LEE, K. H., K. CHUNG, J. M. CHUNG, and R. E. COGGESHALL (1986) Correlation of cell body size, axon size, and signal conduction velocity for individually labelled dorsal root ganglion cells in the cat. J. Comp. Neurol. *243*: 335–346.

LIANG, Y.-F., and S. TERASHIMA (1993) Physiological properties and morphological characteristics of cutaneous and mucosal mechanical nociceptive neurons with A-delta peripheral axons in the trigeminal ganglia of crotaline snakes. J. Comp. Neurol. *328*: 88–102.

MARTIN, A. R., and W. O. WICKELGREN (1971) Sensory cells in the spinal cord of the sea lamprey. J. Physiol. (Lond.) *212*: 65–83.

MATTHEWS, G., and W. O. WICKELGREN (1978) Trigeminal sensory neurons of the sea lamprey. J. Comp. Physiol. *123*: 329–333.

MATSUDA, Y., S. YOSHIDA, and T. YONEZAWA (1978) Tetrodotoxin sensitivity and Ca component of action potentials of mouse dorsal root ganglion cells cultured *in vitro*. Brain Res. *154*: 69–82.

MOLENAAR, G. J. (1978a) The sensory trigeminal system of a snake in the possession of infrared receptors: I. The sensory trigeminal nuclei. J. Comp. Neurol. *179*: 123–136.

MOLENAAR, G. J. (1978b) The sensory trigeminal system of a snake in the possession of infrared receptors: II. The central projections of the trigeminal nerve. J. Comp. Neurol. *179*: 137–152.

PHELAN, K. D., and W. M. FALLS (1989) The interstitial system of the spinal trigeminal tract in the rat: Anatomical evidence for morphological and functional heterogeneity. Somatosens. Mot. Res. *6*: 367–399.

PROSKE, U. (1969a) An electrophysiological analysis of cutaneous mechanoreceptors in a snake. Comp. Biochem. Physiol. *29*: 1039–1045.

PROSKE, U. (1969b) Vibration-sensitive mechanoreceptors in snake skin. Exp. Neurol. *23*: 187–194.

RANSON, S. W. (1912) The structure of the spinal ganglia and of the spinal nerves. J. Comp. Neurol. *22*: 159–175.

RANSON, S. W., and H. K. DAVENPORT (1931) Sensory unmyelinated fibers in the spinal nerves. Am. J. Anat. *48*: 331–353.

RENEHAN, W. E., M. F. JACQUIN, R. D. MOONEY, R. D. RHOADES, and R. W. RHOADES (1986) Structure–function relationships in rat medullary and cervical dorsal horns: II. Medullary dorsal horn cells. J. Neurophysiol. *55*: 1187–1201.

ROSE, R. D., H. R. KOERBER, M. J. SEDIVEC, and L. M. MENDELL (1986) Somal action potential duration differs in identified primary afferents. Neurosci. Lett. *63*: 259–264.

SHIGENAGA, Y., K. OTANI, and S. SUEMUNE (1990) Morphology of central terminations of low-threshold trigeminal primary afferents from facial skin in the cat: Intra-axonal staining with HRP. Brain Res. *523*: 23–50.

SIMINOFF, R., and L. KRUGER (1968) Properties of reptilian cutaneous mechanoreceptors. Exp. Neurol. *20*: 403–414.

STONEY, S. D., JR. (1990) Limitation on impulse conduction at the branch point of afferent axons in frog dorsal root ganglion. Exp. Brain Res. *80*: 512–524.

TALBOT, W. H., I. DARIAN-SMITH, H. H. KORNHUBER, and V. B. MOUNTCASTLE (1968) The sense of flutter-vibration: Comparison of the human capacity with response patterns of mechanoreceptive afferents from the monkey hand. J. Neurophysiol. *31*: 301–334.

TERASHIMA, S., and Y.-F. LIANG (1991) Temperature neurons in the crotaline trigeminal ganglia. J. Neurophysiol. *66*: 623–634.

TERASHIMA, S., and Y.-F. LIANG (1993) Modality difference in the physiological properties and morphological characteristics of the trigeminal sensory neurons. Japan. J. Physiol. *43*: (Suppl. 1): S267–S274.

TERASHIMA, S., Y.-F. LIANG, and Z.-Z. HUANG (1989) Classification of the sensory neurons in the crotaline trigeminal ganglia. XXXI Int. Union Physiol. Sci. Abstr. *17*: 203.

TSURU, K., K. OTANI, S. KAJIYAMA, S. SUEMUNE, and Y. SHIGENAGA (1989) Central termination of periodontal mechanoreceptive and tooth pulp afferents in the trigeminal principal and oral nuclei of the cat. Brain Res. *485*: 29–61.

WINDLE, W. F. (1926) Non-bifurcating nerve fibers of the trigeminal nerve. J. Comp. Neurol. *40*: 229–240.

YOSHIDA, S., and Y. MATSUDA (1979) Studies on sensory neurons of the mouse with intracellular-recording and horseradish peroxidase-injection techniques. J. Neurophysiol. *42*: 1134–1145.

YOSHIDA, S., Y. MATSUDA, and A. SAMEJIMA (1978) Tetrodotoxin-resistant sodium and calcium components of action potentials in dorsal root ganglion cells of the adult mouse. J. Neurophysiol. *41*: 1096–1106.

Modality Difference in the Physiological Properties and Morphological Characteristics of the Trigeminal Sensory Neurons

Shin-ichi TERASHIMA and Yun-Fei LIANG

Department of Physiology, University of the Ryukyus School of Medicine, Nishihara-cho, Okinawa, 903-01 Japan

Abstract   A-δ nociceptive neurons divided according to fast-conducting ($n=21$) and slow-conducting ($n=13$) types were found to be different in electrophysiology and morphology, while between warm temperature ($n=23$) and warm temperature and touch ($n=38$) neurons, and between touch ($n=26$) and vibrotactile ($n=59$) neurons their characteristics were found to be similar.

Key words: trigeminal nerve, ganglion neuron, electrophysiology and morphology.

Sensory ganglion neurons are known to vary in size and form (rat [1], human, ass, sheep, dog, and chicken [2]). Although their morphological characteristics are usually classified according to the conduction velocity (rat [3]), their morphological variety has seldom been correlated directly to the sensory modality. Some modality-identified neurons have been known only by their central projection (rat [4]). The electrophysiological properties of ganglion neurons are also not homogeneous (mouse [5]). The electrophysiological properties of mechanical neurons have been correlated according to their modality (cat [6]), but not correlated according to morphology. Correlation to the sensory modality has not been successful because of the difficulty of doing *in vivo* intrasomal recording along with injection of dye. Utilizing suitable experimental animals for intrasomal recording of ganglion neurons *in vivo*, the present experiment was conducted with the aim of correlating the electrophysiological properties and morphological characteristics to the modality. A preliminary report was published elsewhere [7].

Experimental animals were male and female pit vipers, *Trimeresurus flavoviridis*, of 250–450 g in weight. Halothane gas was used for anaesthesia, and then pancuronium (2 mg/kg, IM) was given for immobilization with artificial respiration. The snake's head was fixed with two pairs of snake head holders, and then the ganglion was exposed by drilling through the inner ear. The dura was cut open with a hypodermic needle. For intrasomal recording and HRP-injection of sensory neurons *in vivo*, an HRP-filled KCl micropipette was used. A 4–6% solution of

HRP in 0.5 M KCl and 0.05 M Tris-buffer (pH 7.3) was put into the micropipette with 60–100 MΩ resistance at 1 kHz. The needle was controlled by a manipulator.

The room temperature was kept at 25°C. For modality identification, a forceps either chilled (0°C) or warmed (45–60°C) by water was used for temperature stimuli. An He-Ne laser (2 mW) was used for detecting the receptive fields (RFs) of warm neuron. A series of von Frey hair was used for determining mechanical thresholds (5, 10, 20, ... 100, 200, ... 1,000, 2,000, ... 5,000 mg), and pin pricks were utilized as a nociceptive stimulus (more than 5 g). A hand-held vibrator made from a dynamic speaker was used to give sinusoidal vibration from 1 to 500 Hz.

Intrasomal potentials were amplified and were observed on a memory oscilloscope and at the same time were stored on the magnetic tape of a data recorder (Sony, band width DC to 10 kHz) for measurement of action potential parameters. The superficial branch of the maxillary division of the trigeminal nerve was exposed just under the eye and was placed on the stimulus electrodes [8] for measurement of conduction velocity (CV) of peripheral axons. The latency from stimulus time to the moment when the elicited action potential arrived at the soma was measured with an intrasomal electrode. For calculation of the CV, the distance of the course was divided by the latency, which was measured along the nerve after the animal was sacrificed.

A positive current was passed through an HRP-filled electrode for HRP intrasomal injection by electrophoresis. The current was repeatedly alternated for a 15–20 min period between being on for 10–30 nA for 1,500 ms and off for 500 ms.

For morphological investigation, the brain was perfused through the heart via the carotid arteries of both sides. The perfusion was first done with heparinized saline (7 IU/ml), next with fixative (2% paraformaldehyde and 2.5% glutaraldehyde in 0.1 M phosphate-buffer, pH 7.4, 10°C), and finally with a phosphate buffer with 10% sucrose (PBS); each perfusion was for more than 20 min. The brain and the trigeminal ganglia of both sides were then removed and kept at 4°C in PBS over night. Frozen serial sections were prepared at a thickness of 80 μm. Horizontal sections of the ganglion and of the brain stem were reacted with nickel-cobalt diaminobenzidine (DAB) [9], then mounted on a glass slide and counterstained with cresyl violet. Neurons were observed under a light microscope (Olympus BH-2). The camera lucida technique was employed to delineate a whole neuron. Our morphological nomenclature followed that of Molenaar (snake [10, 11]). Cochran-Cox's two-tailed t test was used to determine statistical significance [12].

The 180 neurons recorded from 108 pit vipers included 23 warm temperature (T) and 38 warm temperature and touch (T+M) neurons, 26 touch (M) and 59 vibrotactile (V+M) neurons, and 34 A-δ nociceptive neurons which were divided into 21 fast-conducting (FC) and 13 slow-conducting (SC) types.

Electrophysiological properties, such as the resting membrane potential (MP), action potential (AP) parameters, and CV (Table 1) and morphological characteristics, such as axon diameter and soma diameter (Table 2), and central projection (Table 3), were examined. Soma diameters were almost identical for six modalities

Table 1. Physiological properties.

	M	V+M	Warm T	Warm T+M	FC	SC
AP amplitude, mV	68.08±8.29(26)	77.8±11.5(59)	63.48±12.30(23)	61.93±11.26(38)	83.9±13.0(21)	91.8±11.3(13)
AP overshoot beyond the 0-mV level, mV	11.07±6.16(26)	14.3±7.4(55)	8.35±4.92(23)	8.84±4.89(38)	21.0±7.4(21)	27.3±5.7(13)
AP duration at resting potential level, ms	1.68±0.23(16)	1.3±0.2(29)	2.48±0.47(14)	2.56±0.59(22)	2.4±0.3(16)	4.1±0.8(11)
Time to peak, ms	0.64±0.10(16)	0.5±0.1(29)	0.96±0.16(14)	0.93±0.11(22)	1.0±0.2(16)	1.7±0.4(11)
Peak rate of depolarization, V/s	209.0±29.0(13)	356.8±83.3(28)	138.6±34.4(14)	146.3±28.0(22)	182.5±27.5(16)	118.2±31.0(11)
Peak rate of repolarization, V/s	-102.0±10.0(13)	-175.7±53.9(28)	-73.6±25.0(14)	-76.8±29.7(22)	-85.6±12.7(16)	-53.6±14.3(11)
AHP height, mV	-5.06±1.52(16)	-4.3±2.80(29)	-7.07±2.28(14)	-7.27±3.14(22)	-11.9±3.8(16)	-14.1±4.0(11)
AHP duration to half-decay, ms	1.75±0.35(16)	1.5±0.4(23)	3.81±0.54(14)	4.42±0.75(22)	4.5±1.1(16)	25.6±9.8(11)
Resting membrane potential, mV	57.0±5.1(26)	63.3±7.8(48)	55.13±8.41(23)	53.13±8.30(38)	62.9±6.7(21)	64.0±6.3(13)
Conduction velocity, m/s	28.4±5.7(11)	30.2±7.2(23)	11.5±2.9(15)	11.5±2.4(10)	11.2±2.0(21)	3.8±1.1(13)

Values are means±SD; Numbers in parentheses indicate number of neurons; AP, action potential; AHP, after-hyperpolarization.

Table 2. Morphological characteristics.

	M	V+M	Warm T	Warm T+M	FC	SC
Stem axon, μm	5.67±1.89 (9)	5.0±1.5 (16)	4.00±0.50 (12)	3.95±0.33 (11)	4.0±0.5 (9)	2.7±0.4 (8)
Central axon, μm	5.63±2.06 (8)	4.6±1.9 (15)	2.83±0.51 (12)	2.82±0.39 (11)	2.2±0.5 (9)	0.9±0.2 (8)
Peripheral axon, μm	5.78±1.75 (9)	4.9±1.6 (15)	4.04±0.38 (12)	4.00±0.37 (11)	4.0±0.6 (9)	2.5±0.4 (8)
Major diameter, μm	47.0±12.6 (8)	52.6±10.1 (14)	42.1±5.3 (9)	40.6±3.1 (5)	47.8±6.2 (9)	44.3±6.3 (8)
Minor diameter, μm	27.4±6.6 (8)	37.4±8.2 (14)	31.2±4.7 (9)	26.0±3.1 (5)	36.9±5.4 (9)	33.8±4.2 (8)

Values are means ± SD; Numbers in parentheses indicate number of neurons.

or submodalities mentioned above (Table 2), although some very large soma were found in M and V + M neurons.

All RFs of the warm T and warm T + M neurons were found on the heat sensitive pit membrane. There were two kinds of RF, one of which was sensitive to a mechanical stimulation of 10 mg or less (T + M neuron), and the other to 100 mg or more (T neuron). Both were slow-adapting types of receptors. The RF was one for one T or one for one T + M neuron, with the diameter of each being less than

Table 3.   Central projections (terminal density).

	M	V+M	Warm T	Warm T+M	FC	SC
TPR	+++	+++	−	−	++	−
TTDO	+++	+++	−	−	++	−
TTDI	++	++	−	−	+	+
TTDC	+	+	−	−	+	+
LTTD	−	−	+	+	−	−
C_1–C_2	±	±	−	−	±	±
Int.	−	−	−	−	+	+

TPR, principalis; TTDO, oralis; TTDI, interpolaris; TTDC, caudalis; LTTD, lateral descending nucleus; C_1–C_2, 1st to 2nd cervical segment; Int., interstitial nucleus; +++, dense; ++, moderate; +, sparse; ±, descending fiber only; −, absent.

1 mm (the measurement was not precise [8]).

Warm T and warm T+M neurons were similar in physiological properties (Table 1) and morphological characteristics (Tables 2 and 3). They had APs of medium width and after-hyperpolarizations (AHPs) of medium height and duration (Table 1). Both had medium CVs (A-δ range); and they had both stem and peripheral myelinated axons of medium diameter, with a thin myelinated central axon (Table 2). They projected ipsilaterally only to the lateral descending nucleus of the trigeminal tract (LTTD) (Table 3).

The M neuron was a slow-adapting mechanoreceptor and the V+M neuron was a rapidly-adapting mechanoreceptor. They were not temperature-sensitive. The RF was one for one neuron for each type, and the diameters of the RF were 2.1 mm for M, and 1.5–3.5 mm for V+M. The mechanical threshold was 10 mg or less for each. The M neuron maintained a discharge train during the sustained stimulation, but could not elicit one impulse to one cycle of vibration, or even 1 Hz sinusoidal wave. The V+M neuron responded to steady mechanical stimulation with one spike discharge at the initial moment, but the successive spike discharge of 1 : 1 could be maintained by properly adjusting the stimulus intensity within the range of 5–300 Hz.

The M and V+M neurons had spikes of short duration, and also had AHPs of low height and of short duration (Table 1). Both also had fast CVs (A-β range). Some of the M and V+M neurons were found to have large somata, but the others could not be distinguished from those of other modalities. M and V+M neurons had similar diameters ($p > 0.2$) for the set of 3 myelinated axons (stem, peripheral, and central, Table 2). Neither was there any difference found in the central projection (Table 3), which was limited to the ipsilateral principal sensory nucleus of the trigeminal tract (TPR) and to the 3 subnuclei (TTDO, TTDI, TTDC) of the descending nucleus of the trigeminal tract (TTD).

A-δ nociceptive neurons were classified into two types according to the CV: fast-conducting (FC, 11.2 m/s) and slow-conducting (SC, 3.8 m/s) types (Table 1). The RF was one for one neuron for both, and the sizes were 2.0 mm in diameter for

FC and 2.2 mm for SC. The mechanical threshold for each was 5 g or more, but they had no thermal sensitivity (0 to 45–70°C). They were rapidly-adapting receptors and tended to show sensitization to repeated noxious stimulation.

The amplitude of the AP was larger in the A-δ nociceptive neurons (83.9, 91.8 mV for FC and SC, respectively) than in M and V+M (68.1, 77.8 mV) due to the more prominent overshoot in the A-δ group. Only the SC type had a hump in the falling phase of the AP (cat [13]). Another clear difference between them was the duration of the AP. The FC and SC types had mean AP durations of 2.4 and 4.1 ms, respectively, in contrast to 1.7 and 1.3 ms for A-β M and V+M neurons (Table 1). In both types of neurons spikes were followed by hyperpolarizations of different amplitudes (11.9, 14.1 mV for FC and SC neurons and 5.1, 4.3 mV for M and V+ M neurons). The mean duration of AHP was also different in both groups (4.5, 25.6 ms for FC and SC, 1.8, 1.5 ms for M and V+M neurons).

The FC type had a thin peripheral axon and a thinner central axon (Table 2) (cat [14]), both of which were myelinated. Their central axons sent collaterals ipsilaterally to the TPR and the TTD. Five of the 9 successfully labeled neurons sent collaterals to the interstitial nucleus (Table 3). The SC type had the widest APs, and the highest AHPs of the longest duration. Their CVs were slowest in the six submodalities (Table 1). They each had a myelinated peripheral axon and an unmyelinated central axon (cat [15]). All of these neurons projected to the TTDI and the TTDC, but sent no collateral to the TPR or the TTDO, and 5 of 8 neurons projected to the interstitial nucleus. Both the FC type (2 of 9) and SC type (1 of 8) had a branching of the myelinated peripheral axon in the ganglion.

In earlier observations, the CVs in the fiber of the ganglion neurons were found to be faster in the peripheral axon than in the central axon (cat [16–18]). Although we did not measure the CV in the central axon, the results of our present morphological observation of the central axon of A-δ neurons (warm neurons, FC type, and SC type) are consistent with these earlier observations, but our findings concerning the A-β M and V+M neurons are different.

The functional difference in mechanical sensitivity for warm T and T+M neurons was not reflected in our results concerning soma sizes, axon diameters, and central projections. Therefore, it seems that only temperature information common to both submodalities would be utilized in this system. Their sole projection to the LTTD (Table 3) would indicate that no interference from or to other modalities could occur.

The functional difference of M and V+M neurons such as slow- and rapid-adapting receptors, was not reflected in morphology. This morphological evidence indicates that they are united for processing mechanical information for a larger range. As Talbot et al. (human and monkey [19]) suggested, two distinct sets of primary afferent fibers serve in sensing flutter-vibration, one terminal is sensitive to frequencies of 5–40 Hz, and a second terminal is sensitive over the range of 60–300 Hz, although there is a remarkable change in the sensation experienced.

This is the first time that two submodalities such as types FC and SC have been proven to exist in the A-δ nociceptive neurons, and that they have differences in electrophysiology (Table 1), neuronal morphology (Table 2), and central projection (Table 3). Our results on the SC neuron are consistent with other findings concerning the A-δ high-threshold mechanoreceptors which show spikes of larger amplitude and duration, and of longer AHP duration than the other mechanoreceptors (cat [6]).

Branching of the peripheral axon of the A-δ neuron was reported by Hoheisel and Mense (cat [14]) although the modality was not identified. For A-δ nociceptive neurons their response mode showed no difference as far as we have checked by pin prick, but their electrophysiological and morphological differences were unexpectedly great. Electrophysiological differences were found in AP parameters, and in CV (Table 1). Morphological differences were found in the diameters of a set of three axons, especially in the unmyelinated central axon (Table 2), and the central projection sites (Table 3). These differences seem to be based on some functional difference in information processing.

Both FC and SC types are related to nociceptive sensation. Also the FC type seems to relate to a pain related trigeminal reflex in motor activities [20] or in motivational affective mechanisms [21]. The histology that thickest nociceptive fibers terminate in the sensory nuclei near the trigeminal motor nucleus seems to relate to the motor reflex.

The FC neuron shares the projection sites with M and V+M neurons except for the interstitial nucleus. Although terminal distribution in laminae of these nuclei was not determined, there might be the possibility of interaction (cat [22]) between fast- (M or V+M neuron) and slow-conducting (FC type nociceptive neuron) fibers in these nuclei.

The interstitial nucleus was considered to be a pain-related nucleus from the evidence of HRP transganglionic transport from the cornea (cat [23], rat [24]). Hayashi and Tabata (cat [25]) recorded only nociceptive neurons from the nucleus. Our present results showing that only types FC and SC project to the nucleus are consistent with that view. The caudal part of the trigeminal sensory complex, including the interstitial nucleus, seems to be more important in nociception when compared with the rostral part.

In conclusion, although many researchers (review [26]) have reported the structure of the soma and the axons of sensory neurons in the ganglion, this is the first time the soma and a set of three axons have been revealed along with the central projection of the sensory neuron. The results obtained from this technique throws new light on the classification of submodalities of the ganglion neuron. Even though a simple pin prick could not distinguish the submodalities of the A-δ nociceptive neurons, our data which combined electrophysiological properties with morphological characteristics of the neurons enabled us to divide them more consistently into the two types: fast-conducting and slow-conducting. On the other hand, for two known submodalities, that is, between warm temperature and warm temperature plus touch neurons [8], and between touch and vibrotactile neurons

[19], their electrophysiological and anatomical data were found to be similar. Further study with this technique of combining intrasomal recording with intrasomal injection of dye will help to clarify the sensory mechanism in the peripheral nerves and in the CNS.

REFERENCES

1. Hatai S: Number of size of the spinal ganglion cells and dorsal root fibres in the white rat at different ages. J Comp Neur **12**: 107–124, 1902

2. Cajar SR: Die Struktur der sensiblen Ganglien des Menschen unt der Tiere. Anat Hefte Zweite Abt Bd **16**: 177–215, 1906

3. Harper AA and Lawson SN: Conduction velocity is related to morphological cell type in rat dorsal root ganglion neurons. J Physiol (Lond) **359**: 31–46, 1985

4. Hayashi H: Morphology of central terminations of intraaxonally stained, large, myelinated primary afferent fiber from facial skin in the rat. J Comp Neurol **237**: 195–215, 1985

5. Yoshida S and Matsuda Y: Studies on sensory neurons of the mouse with intracellular-recording and horseradish peroxidase-injection techniques. J Neurophysiol **42**: 1134–1145, 1979

6. Koerber HR, Druzinsky RE, and Mendell LM: Properties of somata of spinal dorsal root ganglion cells differ according to peripheral receptor innervated. J Neurophysiol **60**: 1584–1596, 1988

7. Terashima S and Liang Y-F: Central projections of A-δ nociceptive neurons in the trigeminal ganglion of the crotaline snake, *Trimeresurus flavoviridis.* J Physiol (Lond) **446**: 158, 1992

8. Terashima S and Liang Y-F: Temperature neurons in the crotaline trigeminal ganglia. J Neurophysiol **66**: 623–634, 1991

9. Adams JC: Heavy metal intensification of DAB-based HRP reaction product (Letter). J Histochem Cytochem **29**: 775, 1981

10. Molenaar GJ: The sensory trigeminal system of a snake in the possession of infrared receptors. I. The sensory trigeminal nuclei. J Comp Neurol **179**: 123–136, 1978

11. Molenaar GJ: The sensory trigeminal system of a snake in the possession of infrared receptors. II. The central projections of the trigeminal nerve. J Comp Neurol **179**: 137–152, 1978

12. Cochran WG and Cox GM: Experimental Designs, John Wiley & Sons, Inc., New York, 1957

13. Rose RD, Koerber HR, Sedivec MJ, and Mendell LM: Somal action potential duration differs in identified primary afferents. Neurosci Lett **63**: 259–264, 1986

14. Hoheisel U and Mense S: Observations on the morphology of axons and somata of slowly conducting dorsal root ganglion cells in the cat. Brain Res **423**: 269–278, 1987

15. Ha H: Axonal bifurcation in the dorsal root ganglion of the cat: a light and electron microscopic study. J Comp Neurol **140**: 227–240, 1970

16. Gasser HS: Properties of dorsal root unmedullated fibers on the two sides of the ganglion. J Gen Physiol **38**: 709–728, 1955

17. Darian-Smith I, Mutton P, and Proctor R: Functional organization of tactile cutaneous afferents within the semilunar ganglion and trigeminal spinal tract of the cat. J

Neurophysiol **28**: 682–694, 1965

18. Lee KH, Chung K, Chung JM, and Coggeshall RE: Correlation of the cell body size, and signal conduction velocity for individually labelled dorsal root ganglion cells in the cat. J Comp Neurol **243**: 335–346, 1986

19. Talbot WH, Darian-Smith I, Kornhuber HH, and Mountcastle VB: The sense of flutter-vibration of the human capacity with response patterns of mechanoreceptive afferents from the monkey hand. J Neurophysiol **31**: 301–335, 1968

20. Hu JW and Sessle BJ: Comparison of responses of cutaneous nociceptive and non-nociceptive brain stem neurons in trigeminal subnucleus caudalis (medullary dorsal horn) and subnucleus oralis to natural and electrical stimulation of tooth pulp. J Neurophysiol **52**: 39–53, 1984

21. Melzack R and Casey KL: Sensory, motivational, and central control determinants of pain: a new conceptual model. *In* The Skin Senses, ed. Kenshalo DR, Thomas, Springfield, IL, pp 423–443, 1968

22. Ralston HJ: The organization of the substantia gelatinosa Rolandi in the cat lumbo-sacral spinal cord. Z Zellforsch **67**: 1–23, 1965

23. Shigenaga Y, Chen IC, Suemune S, Nishimori T, Nasution ID, Yoshida A, Sato H, Okamoto T, Sera M, and Hosoi M: Oral and facial representation within the medullary and upper cervical dorsal horns in the cat. J Comp Neurol **243**: 388–408, 1986

24. Marfurt CF and Der Toro DR: Corneal sensory pathway in the rat: a horseradish peroxidase tracing study. J Comp Neurol **261**: 450–459, 1987

25. Hayashi H and Tabata T: Physiological properties of sensory neurons of the interstitial nucleus in the spinal trigeminal tract. Exp Neurol **105**: 219–220, 1989

26. Lieberman AR: Sensory ganglia. *In*: The Peripheral Nerve, ed. Landon DN, Wiley, New York, pp 188–278, 1976

Physiological Properties and Morphological Characteristics of Cutaneous and Mucosal Mechanical Nociceptive Neurons With A-δ Peripheral Axons in the Trigeminal Ganglia of Crotaline Snakes

YUN-FEI LIANG AND SHIN-ICHI TERASHIMA
Department of Physiology, University of the Ryukyus School of Medicine,
Okinawa 903-01, Japan

ABSTRACT

Primary A-δ nociceptive neurons in the trigeminal ganglia of immobilized crotaline snakes were examined by intrasomal recording and injection of horseradish peroxidase in vivo. Thirty-four neurons supplying the oral mucosa or facial skin were identified as A-δ nociceptive neurons which responded exclusively to noxious mechanical stimuli and had a peripheral conduction velocity ranging from 2.6 to 15.4 m/s. These neurons were subdivided into a fast-conducting type (FC-type) and a slowly conducting type (SC-type). Neurons of both types had a receptive field limited to a single spot which responded to pin prick stimulus with a threshold of more than 5 g. The FC-type neurons had a narrow spike followed by a shorter after-hyperpolarization. In contrast, SC-type neurons exhibited a broad spike with a hump on the falling phase and a longer after-hyperpolarization. The diameters of the stem, central and peripheral axons of the FC-type neurons were significantly thicker than those of the SC-type neurons, but there was no statistical difference in the soma size of the two types. Central axons of both types of neurons were thinner than their stem and peripheral axons. Dichotomizing fibers of peripheral axons were observed within the ganglion on 3 neurons. Central axons of the FC-type neurons terminated ipsilaterally in the nucleus principalis, the subnucleus oralis, interpolaris and caudalis and the interstitial nucleus, whereas those of the SC-type neurons generally projected only to the caudal half of the subnucleus interpolaris, subnucleus caudalis and interstitial nucleus ipsilaterally. The present data showed for the first time the physiological and morphological heterogeneity of the primary trigeminal A-δ nociceptive neurons and revealed that the trigeminal nucleus principalis and all the subdivisions of the trigeminal descending nucleus are involved in nociception as relay nuclei, but the subnucleus caudalis and the caudal half subnucleus interpolaris are the essential relay sites of the primary nociceptive afferents supplying the oral mucosa and facial skin. The interstitial nucleus also appears to play an important role in orofacial nociception.

Key words: action potential, central projection, interstitial nucleus, dichotomizing fiber

A substantial proportion of primary afferent neurons have been reliably identified as A-δ mechanical nociceptive neurons or so-called high-threshold mechanoreceptive afferents (A-δ HTMs) on the basis of both their axonal conduction velocity (CV) and their sensory modality in humans and various animals (for reviews, see Burgess and Perl, '73; Dubner and Bennett, '83; Sessle, '89). These neurons have thin myelinated peripheral axons and respond to intense mechanical stimulation of their cutaneous receptive fields (RFs), but not to noxious thermal and chemical stimuli. Knowing their physiological properties and morphological characteristics is essential to the complete understanding of the mechanism of pain. Therefore, investigations concerning such neurons have been extensively performed in the dorsal root ganglion (for reviews, see Perl, '68; Dubner and Bennett, '83). However, the primary A-δ HTMs in the

137

trigeminal ganglion have been much less studied, largely because of the difficulty in approaching the ganglion, which in many kinds of animals is covered by the forebrain.

Some important information has been obtained from modality-identified A-δ HTMs of the facial skin and oral mucosa by means of extracellular recording from the trigeminal ganglion or peripheral single fibers (Dubner and Hu, '77; Hu and Sessle, '88; Cooper et al., '91), but up to the present time, no one has reported intrasomatic recording and labeling of these neurons in vivo, and their detailed physiological and morphological characteristics such as intrasomal action potential properties, soma size and axon size have remained unknown.

Central projections appear to be much more complicated for the trigeminal ganglion cells than for the dorsal root ganglion cells, for sensory information from the orofacial area is primarily sent to the trigeminal sensory complex (TSC), which consists of 4 morphologically and functionally distinct components, namely, the nucleus principalis (TPR), subnucleus oralis (TTDO), subnucleus interpolaris (TTDI), and subnucleus caudalis (TTDC) (Molenaar, '78a,b). Since Sjoqvist ('38) succeeded in the treatment of trigeminal neuralgia by section of the spinal tract slightly rostral to the obex, and Grant, in 1939, made an incision a little below the obex which resulted also in complete analgesia in the trigeminal nerve territory, it has long been suggested that the TTDC is the essential component serving as the brainstem relay site for orofacial pain. Numerous morphological and physiological experiments have subsequently supported this classical view (for reviews, see Dubner et al., '76; Yokota, '85). Although the TTDC appears to be particularly important for orofacial nociception, many investigations have suggested that the rostral part of the TSC is also involved in trigeminal nociception. Clinical trigeminal tractotomy at the level of the obex showed that painful orofacial sensations were partially persistent in a certain percentage of patients with trigeminal neuralgia (McKenzie, '55), or showed persistence of pain evoked by tooth pulp stimulation and only hypoalgesia of the oral mucosa and the skin of the paramedian facial region (Young, '82). Tractotomy performed on animals showed that the orofacial nociceptive reflex or behavioral reactions were maintained after operation (Sumino, '71; Young and Perryman, '84). Nociceptive neurons have also been recorded in the rostral part of the TSC (Sessle and Greenwood, '76; Azerad et al., '82; Hayashi et al., '84; Young and Perryman, '86; Dallel et al., '90).

Observations of transganglionic transport of horseradish peroxidase (HRP) from tooth pulp have revealed that tooth pulp afferent fibers terminate in all subdivisions of the TSC, with the heaviest projections to the TPR and the TTDO (Arvidsson and Gobel, '81; Marfurt and Turner, '84). By means of recording of single unit activities in the thalamus combined with trigeminal tractotomy, it was learned that the rostral part of the TSC receives nociceptive afferents mainly from the oral and perioral areas and serves as a relay in ascending pathways which convey painful sensations (Dallel et al., '88).

Taken all together, it appears that the rostral part of the TSC is also involved in nociception. However, the accuracy of conclusions on nociceptive central projections derived from the above-described studies appears to be influenced by several disadvantageous factors. The first is the extreme difficulty of severing the entire trigeminal spinal tract without injury to adjacent structures such as subnuclei of the TSC, internuclear connections, and other pathways. An incomplete incision of the trigeminal spinal tract or injury to the internuclear connections would disturb the interpretation of the experimental data. These might be responsible for the diversity of the results of tractotomy (Grant, '55). The second is that the observation of neuronal activity may also be affected by the interactions between the caudal and rostral parts of the TSC (Khayyat et al., '75; Jacquin et al., '90) aroused through the ascending and descending internuclear pathways (Stewart and King, '63; Nasution and Shigenaga, '87; Jacquin et al., '90). The third factor is that the recording of unit activity and trigeminal tractotomy techniques can not answer the question whether a given nociceptive neuron projects primarily to only one or to a number of subdivisions of the trigeminal sensory complex. A fourth factor is that the transganglionic transport of HRP from the tooth pulp and other observations concerning the tooth pulp do not totally reflect the nociception because the peripheral nerves supplying the tooth pulp also contain certain fibers which convey only innocuous sensory information (McGrath et al., '83; for more references, see Maciewicz et al., '88). Therefore, more direct approaches are now of great importance for mapping the trigeminal nociceptive central projections. Intrasomal injection of a tracer into physiologically identified primary nociceptive neurons, which has been successfully performed in our previous study of temperature neurons (Terashima and Liang, '91), is a suitable approach to this subject.

In the present study we have defined the intrasomal action potential properties of primary trigeminal A-δ mechanical nociceptive neurons and revealed their morphological characteristics with special focus on the central projections. These results should increase our understanding of the peripheral and central mechanism of trigeminal nociception. Some of the data has been briefly reported in abstract form (Terashima and Liang, '92).

MATERIALS AND METHODS
Animal preparation

Twenty-three crotaline snakes of either sex weighing 250–450 g were anaesthetized with halothane followed by pancuronium immobilization (2 mg/kg). Artificial respiration with an unidirectional airflow was applied by means of an aquarium air pump and the flow through the respiratory system was maintained at 0.5 l/min. A few hypodermic needles inserted into the posterior air sack of the lung served as an outlet for the air. The animals were fixed with two pairs of snake head holders at five points to provide a stable condition for intracellular recording. A hole was drilled in the lateral skull on each side to expose the trigeminal ganglion through the inner ear. In each experiment either the bilateral mandibular branches of the trigeminal nerves were exposed about 10 mm from the trigeminal ganglion or the superficial maxillary branches were exposed just under the eye to permit placement of a pair of bipolar stimulating electrodes for measurements of peripheral axonal CV. A pool was formed on each side by skin flaps and filled with liquid paraffin to moisten the exposed nerves. The heart rate was continuously monitored at 50–90 strokes/min and the room temperature was kept at 24–26°C, which is the optimum temperature for this animal. Supplementary doses of pancuronium were administered when necessary during the experiments.

Recording and stimulation

Glass microelectrodes filled with a 4–6% solution of horseradish peroxidase (HRP, Sigma type VI) in 0.5 M KCI and 0.05 M Tris-buffer (pH 7.3) and having a DC resistance ranging from 60 to 100 MΩ were used for intrasomal recording and HRP injection in the trigeminal ganglion. The microelectrodes were connected to the preamplifier of a conventional recording system by Ag-AgCl wire. Because the trigeminal ganglion neurons, except for the temperature neurons (Terashima and Liang, '91), had no background discharge, a sudden large DC shift was taken as the first indication of an intracellular recording. After penetration of a cell body, intense square-wave pulses with duration of 100–1,000 μs were applied to the mandibular or superficial branches of the maxillary division of the trigeminal nerves to activate the cell and then the intensity of electrical stimulation was adjusted to a level just above the threshold for measuring the response latencies. After a somal action potential was elicited, the peripheral receptive field (RF) was examined by pricking the facial skin and oral mucosa with a syringe needle. The pricking intensity was strong enough to obtain a painful sensation when applied to the skin of the investigator and to perceive a feeling of penetration of the skin when applied to the snake. Once an RF was located by pricking, a series of stimulations, i.e., touching with a set of von Frey hairs, pressing with a blunt probe or a small cotton ball, vibrating with a specially designed hand-held vibrator, warming with a 30–40°C forceps and cooling with an ice cube, were subsequently applied to it in the order named to exclude the neurons which responded to innocuous stimuli. A noxious thermal stimulus was finally applied using a 45–70°C forceps. Only the neurons which responded exclusively to noxious mechanical stimulation and had a membrane potential (Em) greater than −50 mV and an action potential (AP) overshooting the 0-mV level were used for the measurement of somal AP properties. The parameters of the somal AP were measured as in Figure 1. Time-differentiations were recorded by a differential unit (a modified Nihon Kohden SS-1468) for measurement of the peak rate of depolarization and repolarization, and to facilitate the observation of a hump on the falling phase of the somal spike. Electrophysiological data were recorded on an FM tape recorder (Sony, bandwidth 0–10 kHz) for later study.

Histological processing

After identification of an A-δ mechanical nociceptive neuron, HRP was iontophoretically injected into the soma using a positive current of 10–30 nA with 1,500 ms duration and 500 ms off-time for 15–20 min if the penetration was stable. To avoid confusion in tracing labeled axons, only one HRP injection was made in each ganglion. After the last iontophoresis, the animals were maintained under anaesthesia and paralysis for 24–48 h, and then perfused through the heart with 200 ml of saline at 10°C followed by 200 ml of fixative containing 2% paraformaldehyde and 2.5% glutaraldehyde in 0.1 M phosphate-buffer (pH 7.4, 10°C) and 200 ml of 10% phosphate-buffer sucrose solution (pH 7.4, 10°C). The brainstem with the trigeminal ganglion, upper part of the cervical spinal cord, the tectum, and the cerebellum were removed and stored in 30% sucrose solution in phosphate-buffer overnight at 4°C. Serial frozen sections were cut at 80 μm in the horizontal plane and processed with a nickel-cobalt diaminobenzidine reaction

(Adams, '81). The sections were mounted on chrome alum-gelatin slides and counterstained with 1% Cresyl violet after being air dried.

Soma sizes were determined under the microscope (Olympus BH-2) by measuring both the long and short axis of the cell body when the somata were focused in the plane where they had the greatest cross-section area. The diameters of central, peripheral, and stem axons were measured across the outside of the myelin sheaths within the ganglion at 50 μm intervals from the T-bifurcations with an ocular micrometer calibrated with a stage micrometer. Ten measurements were made on peripheral axons and five on central and stem axons and the mean values were taken as their size. The diameters of central axons within the brainstem at the level of the TTDI were also measured in the same manner. Seventeen labeled neurons could be traced to their central terminals and some of them were reconstructed by means of camera lucida drawings. Detailed descriptions of the topographical identification of the morphological structures of the snake mentioned in this paper and the terminologies and abbreviations used to refer to these structures have been given by Molenaar ('78a,b). Mean ±SD was used to indicate all the statistical average values measured in this study and the statistical analysis of the results was carried out using Cochran-Cox's two-tailed t tests.

RESULTS
Physiological data

Thirty-four primary afferents of 23 snakes were identified as mechanical nociceptive neurons which responded exclusively to noxious mechanical stimuli. They had a peripheral CV ranging from 2.6 to 15.4 m/s and so were all classified as A-δ afferents (Maciewicz et al., '88; Terashima and Liang, '91). Of these neurons, 21, which had peripheral axons conducting at 8.8 to 15.4 m/s with a mean of 11.2 m/s (±2.0 S.D.), were subdivided into FC-type and the remaining 13, which had a peripheral CV ranging from 2.6 to 6.4 m/s with a mean of 3.8 m/s (±1.1 S.D.), into SC-type.

Property of action potential. Neurons sampled in this study had no background discharges prior to application of a noxious mechanical stimulation. All these neurons exhibited a stable Em of over −50 mV after penetrations and an overshooting somal AP elicited by electrical stimulation of the trigeminal nerve branch. The mean Em was 62.9 mV (±6.7 S.D., n = 21) for FC-type and 64.0 mV (±6.3 S.D., n = 13) for SC-type neurons. In each neuron the discharges at very low frequency with rapid adaptation were elicited by pricking its RF with a syringe needle. All neurons showed sensitization to repeated stimuli, but were not activated by noxious heat stimulus in the 45–70°C range regardless of the increases in their sensibility.

The mean AP amplitude was 83.9 mV (±13.0 S.D., n = 21) for FC-type, and 91.8 mV (±11.3 S.D., n = 13) for SC-type neuron. Examples of the somal APs of a FC- and a SC-type neurons evoked by stimulation of the mandibular branch of the trigeminal nerve are illustrated in Figure 1, in which measurements of wave-form parameters of a somal AP are shown. The means of each measured parameter of the somal AP were compared between the FC- and SC-type neurons and the statistical results are shown in Table 1. As shown in Figure 1B, all SC-type neurons exhibited an inflection or hump on the repolarizing phase of the somal AP that resulted in a long duration (mean 4.1 ± 0.8 ms, n = 11, measured at the Em level) of the somal spike. These

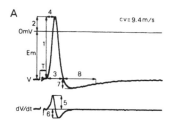

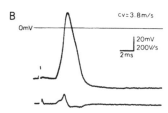

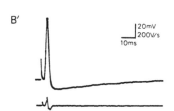

Fig. 1. Somal action potentials (APs) of primary trigeminal A-δ nociceptive neurons responding to electrical stimulation of the mandibular branches of the trigeminal nerve. In each subfigure two records are shown: upper trace, AP; lower trace, time-differentiated response (dV/dt). **A:** An AP of a fast-conducting (FC)-type neuron, demonstrating the wave-form parameters measured as follows: spike latency from electrical stimulus (T), resting membrane potential (Em), AP amplitude (1), AP overshoot beyond the 0-mV level (2), AP duration at the Em level (3), time to peak (4), peak rate of depolarization (5) and repolarization (6), after-hyperpolarization (AHP) height (7), and AHP duration to half-decay to Em (8). **B:** An AP of slowly-conducting (SC)-type neuron, which differs from that of FC-type neurons in parameters 3, 4, 5, 6, and 8 (see Table 1). Note that a hump appears on the falling phase of the spike, which is more clearly seen in the dV/dt record. **B':** A record of the AP in B at slow sweep speed, showing a longer half-decay of AHP (29 ms). cv, conduction velocity of the peripheral axon. Calibration in B applies also to A.

TABLE 1. Electrophysiological Parameters of A-δ Nociceptive Neurons[1]

Measurements	FC-type	SC-type	P
CV of peripheral axon (m/s)	11.2 ± 2.0 (21)	3.8 ± 1.1 (13)	<0.001
Resting membrane potential (mV)	62.9 ± 6.7 (21)	64.0 ± 6.3 (13)	>0.5
AP amplitude (mV)	83.9 ± 13.0 (21)	91.8 ± 11.3 (13)	>0.05
AP overshoot beyond 0-mV level (mV)	21.0 ± 7.4 (21)	27.8 ± 5.7 (13)	<0.02
AP duration at resting potential level (ms)	2.4 ± 0.3 (16)	4.1 ± 0.8 (11)	<0.001
Time to peak (ms)	1.0 ± 0.2 (16)	1.7 ± 0.4 (11)	<0.002
Peak rate of depolarization (V/s)	182.5 ± 27.5 (16)	118.2 ± 31.0 (11)	<0.001
Peak rate of repolarization (V/s)	85.6 ± 12.7 (16)	53.6 ± 14.3 (11)	<0.001
AHP height (mV)	11.9 ± 3.8 (16)	14.1 ± 4.0 (11)	>0.1
AHP duration to half-decay (ms)	4.5 ± 1.1 (16)	25.6 ± 9.8 (11)	<0.001

[1]Probability values for the null hypotheses were tested by the Cochran-Cox's 2-tailed *t* test. Values are means ± S.D. Numerals in parentheses indicate the number of neurons. CV, conduction velocity; AP, action potential; AHP, after-hyperpolarization.

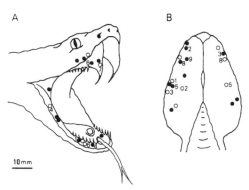

Fig. 2. Receptive fields (RFs) of FC-type (●) and SC-type (○) A-δ mechanical nociceptive neurons. **A:** RFs on the oral mucosa (the uppermost 3 located on the facial skin). **B:** RFs on the skin of the lower jaw. Numbers indicate the RFs of labeled neurons and correspond to those of Table 3 in which the central projection sites of the labeled neurons are summarized.

humps were clearly reflected in the time-differentiated record. The mean half recovery time (half-decay) of after-hyperpolarization (AHP) in SC-type neurons was 25.6 ± 9.8 ms (n = 11). In contrast, no FC-type neurons had any hump on the falling phase of their spikes (Fig. 1A), and therefore, they had a narrow somal spike (mean duration 2.4 ± 0.3 ms, n = 16) followed by a shorter AHP (half-decay 4.5 ± 1.1 ms, n = 16).

Features of receptive field. All neurons of both FC- and SC-types had similar RFs (Fig. 2), which were located on

the facial skin or oral mucosa and limited to a single spot whose mean diameter was 2.0 mm (±0.5 S.D., n = 21) for FC-type and 2.2 mm (±0.5 S.D., n = 13) for SC-type neurons. Of the 21 FC-type neurons, 13 had an RF on the perioral skin and 8 on the oral mucosa. Similarly, 9 out of 13 SC-type neurons had an RF on the perioral skin and the other 4 on the oral mucosa. The RFs were small and homogeneous in sensitivity within the receptive area. Most of the RFs located on the facial skin were restricted within one scale of snake, and a few of them covered two scales where a response could also be evoked by pricking of the interscale region. Only one RF could be found for each neuron in this study. Neurons of both types had also similar mechanical thresholds, which were all over 5g. When an RF was found to be located on the skin by pricking or pinching, a 5g force von Frey hair was applied to it to examine its mechanical threshold. Since the mucosa is very easily damaged by von Frey hairs, when an RF was located on the oral mucosa by pricking, a thin (1 mm in diameter) cotton bar exerting 5g force was then applied to rule out a pressure cell. If a neuron failed to respond to a stimulus of 5g, it was classified as a specific nociceptive neuron.

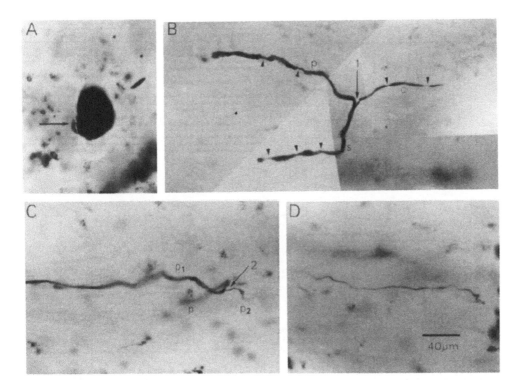

Fig. 3. Photomicrographs taken from the horizontal sections of intracellularly HRP-labeled soma and stem, central and peripheral axons of an FC-type A-δ primary trigeminal nociceptive neuron having a peripheral CV of 10.7 m/s. **A:** Soma with initial segment of stem axon (arrow). **B:** Composite photomicrograph of stem axon (s) and its bifurcation (arrow 1), central axon (c) and peripheral axon (p). Note that the central axon is thinner than the stem and peripheral axons and that all the stem, central, and peripheral axons show the nodes of Ranvier (arrowheads). **C:** Peripheral axon (p) and its bifurcation (arrow 2) from which a myelinated (p1) and an unmyelinated (p2) secondary peripheral axon originates. **D:** A continuation of p2 in C. A camera lucida drawing of the same neuron is shown in Figure 4. Calibration in D applies also to A–C.

Morphological data

Intrasomal injections of HRP were made in 28 physiologically identified A-δ mechanical nociceptive neurons, 17 (61%) of which (9 FC- and 8 SC-type neurons) were labeled well enough to be traced to their central terminals, but their peripheral axons were not well stained after they left the ganglion for a certain distance. In these 17 neurons, however, the cell bodies, stem axons, peripheral axons in the ganglion, central axons, and terminal arbors and boutons were well visualized.

Cell body and axons. The labeled neurons had round or oval soma whose mean long and short diameters were $47.8 \pm 6.2 \times 36.9 \pm 5.4$ μm (n = 9) for FC-type neurons and $44.3 \pm 6.3 \times 33.8 \pm 4.2$ μm (n = 8) for SC-type neurons (Figs. 3–6). There was no difference between the mean somal diameters for both types ($P > 0.5$). The soma sizes of these neurons were medium and fell into a peak area of the distribution in the histogram of soma size of trigeminal ganglion cells in snakes (Terashima and Liang, '91).

Table 2 shows that the mean diameters of the stem, peripheral, and central axons of FC-type neurons were 4.0 ± 0.5 μm (ranging 3.2–4.8 μm, n = 9), 4.0 ± 0.6 μm (ranging 3.2–5.0 μm, n = 9) and 2.2 ± 0.5 μm (ranging 1.5–3.4 μm, n = 9), respectively, and those of SC-type neurons were 2.7 ± 0.4 μm (ranging 2.2–3.4 μm, n = 8), 2.5 ± 0.4 μm (ranging 1.8–3.2 μm, n = 8) and 0.9 ± 0.2 μm (ranging 0.6–1.2 μm, n = 8), respectively. The mean diameters of the stem, central, and peripheral axons in each type and those of the homonymous axons between the two types were compared and the statistical results given in Table 2. In spite of a little overlap between the axon size ranges, diameters of the 3 kinds of axons in FC-types were significantly thicker than those in SC-types. In each type both the stem and peripheral axons were thicker than the central axon, but there was no difference in diameter between them. The central axons of all SC-type neurons were unmyelinated as judged from the absence of nodes of Ranvier (Fig. 5B, 6), but the other axons were all myelin-

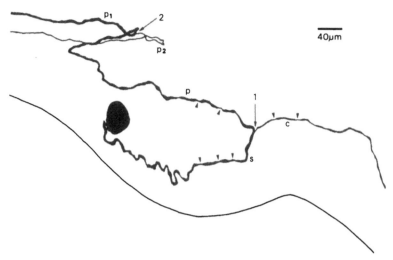

Fig. 4. Camera lucida drawing of the FC-type A-δ nociceptive neuron shown photomicrographically in Figure 3. The curved line at the bottom indicates the lateral margin of the trigeminal ganglion. All the labels correspond to those in Figure 3. Note that the myelinated stem axon (s) bifurcates into a myelinated central axon (c) and a myelinated peripheral axon (p) which dichotomizes into a myelinated (p1) and an unmyelinated (p2) secondary peripheral axon.

ated, and interruptions of the myelin sheaths (i.e., nodes of Ranvier) were clearly observed (Figs. 3, 5). Although the stem axon meandered throughout its course, we could not observe any glomerular structure in which the initial segment of the stem axon winds repeatedly around its neuronal soma (Spencer et al., '73).

Two of 9 FC- and 1 of 8 SC-type neurons gave rise to 2 peripheral axons within the trigeminal ganglion. All these neurons exhibited the same branching pattern, in which a stem axon originating from the soma divided into a central and a peripheral axon and the latter traveled for a distance of several hundred micrometers and then dichotomized into two secondary peripheral axons (Figs. 5, 6). The secondary peripheral axons of one FC- and one SC-type neuron were all thin and myelinated, while the remaining FC-type neuron had one thin myelinated and one unmyelinated peripheral axon. Although it is possible that these neurons may possess two separate RFs, for each of them only one could be located. Also, only one AP could be evoked at a fixed latency when they responded to electrical stimulation of the peripheral nerve trunk. No branching was observed for any labeled central axons within the trigeminal ganglion and root.

Central projections. All A-δ mechanical nociceptive neurons in this study terminated exclusively in the ipsilateral brainstem. The central projecting loci of each labeled neuron are given in Table 3.

Upon entering the pons, the central axon of all FC-type neurons bifurcated into a short ascending and a long descending fiber (Fig. 7A). The ascending fiber was much thinner than the descending fiber (Fig. 7A,B). All ascending fibers gave rise to a number of terminal arbors and boutons which were restricted within the TPR. The descending fibers traveled within the trigeminal spinal tract (ttd) to the

first cervical spinal segment and on the way they gave rise perpendicularly to collaterals at regular intervals to the TTDO, TTDI, and TTDC, where each collateral produced many terminal arbors and boutons (Figs. 7, 8). Five of 9 FC-type neurons gave rise to several very short collaterals restricted within the ttd from caudal TTDI to rostral TTDC. These collaterals produced a small number of the thin terminal arbors and boutons in the interstitial nucleus (Figs. 7E, 8), in which a few small neurons (8–12 μm in diameter) could be observed in the scattering cell pockets. The terminal arbors and boutons were thinner in the caudal half TTDI, TTDC, and interstitial nucleus than those in the TPR, TTDO, and rostral TTDI, and were the most dense in the TPR and TTDO, moderately dense in the TTDI and TTDC, and sparse in the interstitial nucleus. The collaterals and their arbors in the TPR and all the subnuclei of the trigeminal descending nucleus were relatively long, but those in the interstitial nucleus were obviously short (Fig. 8). Since the interstitial nucleus has a shorter distance from the cell body than the TTDC, the low density of the terminal arbors and boutons, and the short length of the collaterals in this nucleus is unlikely due to an incomplete staining of HRP. In the TPR, TTDO, and TTDI, generally, the adjacent arbors did not overlap except when two neighbouring collaterals arising from the descending fiber were close to each other (Fig. 8). No overlapping of arbors was observed in the TTDC and interstitial nucleus. Although all descending fibers could be traced to the first cervical spinal segment, no collaterals, arbors, or boutons were found there. The central axons tapered from a mean diameter of 2.2 ± 0.5 μm in the ganglion to a mean diameter of 1.2 ± 0.3 μm in the ttd at the obex level. The taperings were clearly observed from the entering zone in the pons to the first cervical segment, but were not observed

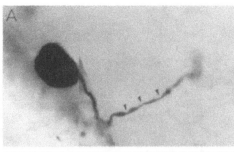

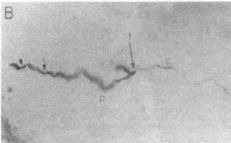

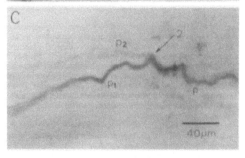

Fig. 5. Photomicrographs taken from the horizontal sections of intracellularly HRP-labeled soma, and stem, central and peripheral axons of an SC-type A-δ primary trigeminal nociceptive neuron having a peripheral CV of 4.6 m/s. **A:** Soma with stem axon. **B:** Stem axon (s) and its bifurcation (arrow 1), central axon (c) and peripheral axon (p). **C:** Peripheral axon (p) and its bifurcation (arrow 2) from which two myelinated secondary peripheral axons (p1 and p2) originate. Note that the stem and peripheral axons have the nodes of Ranvier (arrowheads), while the central axon is a very thin unmyelinated fiber. A camera lucida drawing of the same neuron is shown in Figure 6. Calibration in C applies also to A and B.

in the trigeminal root. A typical example of the central axon and its terminals of an FC-type neuron is illustrated in Figure 8.

Of the 8 labeled SC-type neurons, 7 had no ascending fibers at the entrance to the pons (Fig. 9A) and their thin central axons descended in the ttd as far as the first cervical segment without giving rise to any collaterals until they reached the level of the caudal TTDI. These descending fibers terminated in the caudal half of the TTDI and TTDC. The TPR, TTDO, and the rostral half of the TTDI were totally devoid of the labeled terminal collaterals, arbors, and boutons (Fig. 10). The caudal half of the TTDC also lacked labeled terminals in 3 neurons. Also, five of SC-type neurons were found to terminate in the interstitial nucleus. The arborizations in the caudal TTDI, the TTDC, and the interstitial nucleus of SC-type neurons is very similar to those in the corresponding regions of FC-type neurons; but the collaterals, arbors, and boutons were thinner (Fig. 9B,C). One of the SC-type neuron projected to all the parts of the TSC and interstitial nucleus as FC-type neurons (Table 3); but its terminal collaterals, arbors, and boutons were much thinner. A typical example of the central projections of an SC-type neuron is illustrated in Figure 10. The detailed somatotopical organization of the central projections was not examined in this study.

DISCUSSION
Peripheral conduction velocity and receptive fields

Neurons sampled here had peripheral CVs ranging from 2.6 to 15.4 m/s, which fell into the range of A-δ primary neurons. Our data were very similar to those of Dubner and Hu ('77), who found an average CV of 10.2 m/s (S.D. = 7.3) for A-δ mechanical nociceptive neurons in monkey trigeminal ganglion and to those of Lynn and Carpenter ('82) who reported that most of such neurons with axons innervating rat hind limbs had peripheral CVs of less than 15 m/s. Therefore, it appears that our data may be comparable with those obtained from mammals. All labeled neurons in this study had thin peripheral axons on which nodes of Ranvier could be observed. The labeled peripheral axons had diameters ranging from 1.8 to 5.0 μm corresponding to the range of 2.6 to 14.5 m/s of CVs. This is the first case that CVs were brought into correspondence with axon diameters for identified A-δ mechanical nociceptive neurons. This result indicates that when a peripheral axon of a nociceptive neuron conducts at a velocity of more than 2.5 m/s, it is usually myelinated.

One of the views regarding to primary A-δ mechanical nociceptive neurons is that their RFs consist of many sensitive spots which are separated by insensitive skin areas (for a review, see Burgess and Perl, '73). This concept was mainly based on the experiments carried on the limbs of cats by Burgess and Perl ('67) and on the limbs of monkeys by Perl ('68). They reported that the mechanically sensitive RFs of all such neurons are made up of 3 to 20 excitable small spots (< 1 mm in diameter) arranged in a roughly circular or oval area from under 10 mm to over 20 mm in the longest dimension and separated by areas unresponsive to identical stimuli. These observations were subsequently confirmed on the rat's limbs (Lynn and Carpenter, '82; Martin et al., '87) and the rat's tail (Handwerker et al., '87, Reeh et al., '87). With the electron microscope, Kruger et al. ('81) showed that the fine structure of A-δ mechanical nociceptors ending in the cat's limb skin conforms to the observations mentioned above. On the facial skin and oral mucosa, however, the features of the RFs of A-δ mechanical nociceptive neurons appear to be different from those on limbs. Dubner and Hu ('77) found that such neurons in the trigeminal ganglion of the monkey had an RF on the facial skin limited to a single spot (1–2 mm in diameter). Subsequently, a similar result obtained

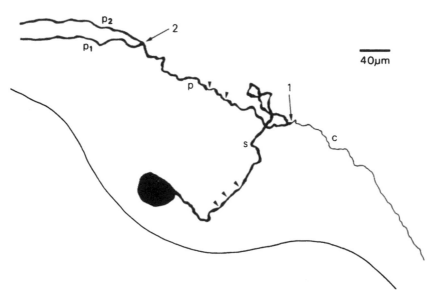

Fig. 6. Camera lucida drawing of the SC-type A-δ nociceptive neuron shown photomicrographically in Figure 5. The curved line at the bottom indicates the lateral margin of the trigeminal ganglion. All the labels correspond to those in Figure 5. Note that the myelinated stem axon (s) bifurcates into an unmyelinated central axon (c) and a myelinated peripheral axon (p) which dichotomizes into two myelinated secondary peripheral axons (P1 and P2).

TABLE 2. Axons of A-δ Nociceptive Neurons[1]

Measurements	FC-type	SC-type	P
Axon diameters (μm)			
Stem axon	4.0 ± 0.5 (9)	2.7 ± 0.4 (8)	<0.001
Central axon	2.2 ± 0.5 (9)	0.9 ± 0.2 (8)	<0.001
Peripheral axon	4.0 ± 0.6 (9)	2.5 ± 0.4 (8)	<0.001
P values			
Stem: central	<0.001	<0.001	
Stem: peripheral	>0.8	>0.2	
Peripheral: central	<0.001	<0.001	

[1]Probability values for the null hypotheses were tested by the Cochran-Cox's 2-tailed t test. Values are means ± S.D. Numerals in parentheses indicate the number of neurons.

from cat's facial skin was reported by Hu and Sessle ('88), who stated that most RFs of A-δ mechanical nociceptive neurons were localized to only one small spot and the remaining few had only two or three spots. Observation on the goat oral mucosa showed that all RFs of A-δ mechanical nociceptive neurons were single, small (1–2 mm in diameter) spots (Cooper et al., '91). In common with these observations, all such neurons encountered in the present study had also an RF limited to a single, small (<2.5 mm in diameter) spot on the facial skin or oral mucosa. Demonstration of a multiple punctate RF requires that the skin be undamaged and carefully examined with small probes. However, it is unlikely that the absence of such an RF on the orofacial area of snake in this study was due to oversight or mistake in the experiments because very fine probes made of syringe or sewing needles with a tip <0.2 mm in diameter were used to examine the RFs with great caution and the skin of snake is not so easily damaged as in most other animals. Therefore, it appears that the RF of

A-δ mechanical nociceptive neurons is smaller in the orofacial area than on the limbs and is usually limited to one spot. This nature indicates that in mechanical nociceptive sensation provoked by activity in A-δ fibers the orofacial territory might have more accurate localization and higher discrimination of the receptive area than the limb.

Diversity of the somal action potential

Intrasomal recordings from the primary neurons in the peripheral ganglion of hamsters (Gurtu and Smith, '88), frog (Stoney, '85), pigeons (Gorke and Pierau, '80), rats (Harper and Lawson, '85), and cats (Rose et al., '86; Koerber et al., '88; Traub and Mendell, '88) have revealed modality dependent differences in the shape of the somal APs and showed varieties in the somal spike width and AHP duration of A-δ neurons. Rose et al. ('86) and Traub and Mendell ('88) reported that in the dorsal root ganglion of the cat all A-δ afferents responsive to noxious mechanical stimulation exhibited a hump on the falling phase of the somal AP that prolongs the duration of the somal spike, whereas all A-δ cells excited by low-threshold natural stimulation had a narrow somal spike without any inflection on the repolarizing phase. Therefore, they suggest that these two categories of neurons in the cat can be identified reliably on the basis of their somal spike shape if the modality is unknown. In contrast to those observations, the present results showed that the primary A-δ mechanical nociceptive neurons with peripheral axons conducting at slower velocity had a hump on the repolarizing phase of the somal spike, but those conducting at faster velocity did not manifest any inflection on the falling phase although their

TABLE 3. Central Projections of A-δ Nociceptive Neurons[1]

Nuclei	No. of FC-type neurons									No. of SC-type neurons							
	1	2	3	4	5	6	7	8	9	1	2	3	4	5	6	7	8
TPR	+++	+++	+++	+++	+++	+++	+++	+++	+++	−	++	−	−	−	−	−	−
TTDO	+++	+++	+++	+++	+++	+++	+++	+++	+++	−	++	−	−	−	−	−	−
TTDIr	++	++	++	++	++	++	++	++	++	−	++	−	−	−	−	−	−
TTDIc	++	++	++	++	++	++	++	++	++	++	++	++	++	++	++	++	++
TTDC	++	++	++	++	++	++	++	++	++	++	++	++	++	++	++	++	++
Int.	−	+	+	+	+	+	−	−	−	−	+	−	+	+	−	+	+

[1]Plus signs indicate the presence and density of boutons and/or arbors in the nuclei. +++, dense; ++, moderate; +, sparse; −, absence; Int., interstitial nucleus; TPR, TTDO, TTDI, and TTDC as in Figure 7; TTDIr, rostral-half TTDI; TTDIc, caudal-half TTDI.

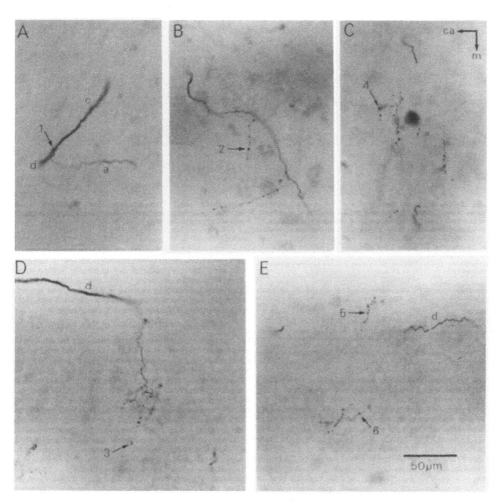

Fig. 7. Photomicrographs taken from the horizontal sections of intracellularly HRP-labeled terminal arbors and boutons of central axon collaterals of an FC-type A-δ primary trigeminal nociceptive neuron having a peripheral CV of 10 m/s. **A:** Central axon (c) and its bifurcation (arrow 1) at the entrance into the pons. It dichotomizes into a thinner ascending fiber (a) and a thicker descending fiber (d). **B:** Terminals in the nucleus principalis. **C:** Terminals in the subnucleus oralis. **D:** Terminals in the subnucleus interpolaris. d, descending fiber in trigeminal spinal tract. **E:** Terminals in the interstitial nucleus (arrow 5) and subnucleus caudalis (arrow 6). A camera lucida drawing of the same axon and its terminals is shown in Figure 8. Calibration in E applies also to all others. ca, caudal; d, descending fiber in trigeminal spinal tract at the obex level; m, medial.

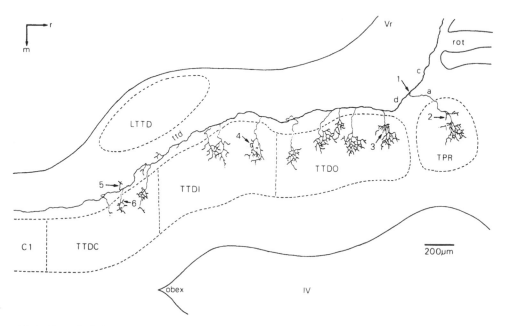

Fig. 8. Camera lucida drawing from horizontal sections of central axon and terminals of the FC-type A-δ nociceptive neuron shown photomicrographically in Figure 7. LTTD, nucleus descendens lateralis; TPR, nucleus principalis; TTDO, subnucleus oralis; TTDI, subnucleus interpolaris; TTDC, subnucleus caudalis; C1, first cervical dorsal horn; ttd, trigeminal spinal tract; Vr, trigeminal root (mandibular and maxillary); rot, radix ophthalmicus; IV, fourth ventricule; r, rostral; m, medial. Arrows 1–6 and a, c, and d correspond to those in Figure 7. Arrow 5 indicates one of three collaterals with arbors and boutons in the interstitial nucleus at the obex level in the ttd. Note that this neuron projects to all the parts of the TSC and interstitial nucleus and its descending fiber extends beyond the TTDC, but no collaterals, arbors or boutons were found in C1.

spike durations at the base-line were as long (> 2 ms) as the spike duration of the A-δ nociceptive neurons showing a hump on the falling phase (Rose et al., '86; Traub and Mendell, '88). The somal APs of FC-type neurons were extremely similar to those of temperature neurons (Terashima and Liang, '91) in all AP parameters. On the other hand, Belmonte and Gallego ('83) reported that in the petrosal ganglion of the cat the baroreceptor neurons responsive to mechanical stimulations of the carotid sinus could be divided on the basis of peripheral CV into two distinct groups: fast neurons which showed no hump on the falling phase of the somal AP and slow neurons with a clear hump on the falling phase of the somal spike. Therefore, it is evident that the properties of the somal AP are not a reliable factor to identify the modality of primary neurons.

Morphological characteristics of the soma and axons

The present study showed that there was no statistical difference in the soma size between FC- and SC-type groups of A-δ nociceptive neurons, although the peripheral CV and axon size were significantly different. Therefore, it is evident that there is no value in knowing the size of an A-δ cell for predicting its peripheral axon size and CV.

A generally accepted opinion is that the central and peripheral axons of the ganglion cells having myelinated fibers are of equal size, whereas in cells having unmyelinated fibers the central axons are thinner than the peripheral axons (Gasser, '55; Lieberman, '76). Recent investigations, however, have provided evidence to the contrary. By means of measuring the transverse sections of the axons in the peripheral nerve and dorsal root in the rat at the electron microscope level, Suh et al. ('84) found that the mean cross sectional areas of central myelinated axons were significantly less than those of peripheral myelinated axons. Thus they suggested that the myelinated central axons of dorsal root ganglion cells are thinner than their myelinated peripheral axons. This suggestion was recently confirmed by Lee and his co-workers ('86) and Hoheisel and Mense ('87), using intracellular staining with HRP in the cat dorsal root ganglion. They showed that the primary A-δ and C cells have thinner central axons than peripheral axons. For the trigeminal ganglion our previous (Terashima and Liang, '91) and present data revealed for the first time that the central axons of A-δ temperature neurons and mechanical nociceptive neurons are significantly thinner than peripheral axons. Our unpublished data showed no difference in size between central and peripheral axons for primary trigeminal neurons with thick myelinated axons conducting at a velocity in the A-β range (no A-α cells were encountered). We concluded, therefore, that the A-δ neurons in the trigeminal ganglia usually give rise to thicker

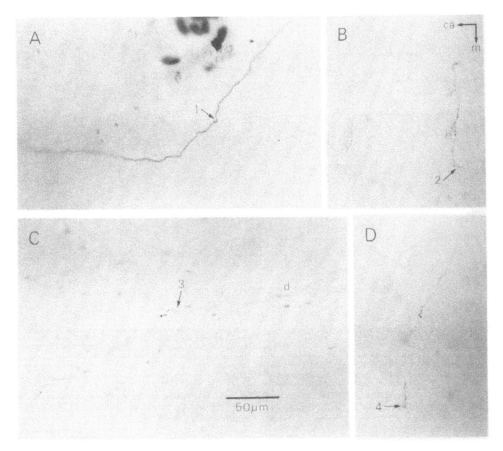

Fig. 9. Photomicrographs taken from the horizontal sections of intracellularly HRP-labeled terminal arbors and boutons of central axon collaterals of an SC-type A-δ primary trigeminal nociceptive neuron having a peripheral CV of 3.6 m/s. **A:** Central axon at the entrance into the pons. Note that there are no ascending fibers given off from the central axon. **B:** Terminals in the subnucleus interpolaris. **C:** Terminals in the interstitial nucleus (arrow 3). **D:** Terminals in the subnucleus caudalis. A camera lucida drawing of the same axon and its terminals is shown in Figure 10. Calibration in C applies also to all others. ca, caudal; d, descending fiber in the trigeminal spinal tract at the obex level; m, medial.

peripheral and thinner central axons, as in the dorsal root ganglion. Although the difference in size between central and peripheral axons would be responsible for a notable decrease in root CV of A-δ and C cells (Loeb, '76; Traub and Mendell, '88), the detailed physiological meaning of this is not clear.

Traub and Mendell ('88) found that A-δ cells in the cat dorsal root ganglion exhibited a two-step decrease in CV (i.e., a sequential decrease from the periphery to the dorsal root to the spinal cord), and stated that the second decrease, which occurred upon entering the spinal cord, was much greater for A-δ high-threshold mechanoreceptive afferents than for A-δ low-threshold mechanoreceptive afferents. In correspondence to their finding, we found that a remark-able tapering of the central axons of A-δ nociceptive neurons occurs in the ttd. Accepting the view that the CV varies with the axon size (Hursh, '39; Boyd and Kalu, '79), our finding of taperings of the central axons in the ttd provides a possible explanation for the decrease in CV within the central nervous system.

The finding that 3 (18%) of 17 labeled neurons had dichotomizing peripheral axons within the ganglion is of importance since one of the hypotheses of referred pain was made based on the suggestion that the same sensory neuron may supply skin and deep or visceral structures by branching its peripheral axon into different nerves (Sinclair et al., '48; Bahr et al., '81; Langford and Coggeshall, '81). The existence of peripheral axonal branching in or near the

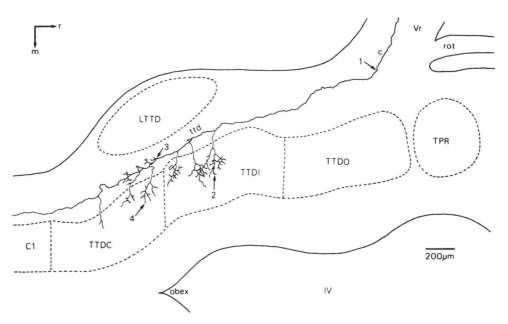

Fig. 10. Camera lucida drawing from horizontal sections of the central axon and terminals of the SC-type A-δ nociceptive neuron shown photomicrographically in Figure 9. Arrows 1–4 correspond to those in Figure 9. Arrow 3 indicates one of several collaterals with arbors and boutons in the interstitial nucleus at the obex level in the ttd. c, central axon. Other abbreviations as in Figure 8. Note that this neuron projects only to the caudal TTDI, TTDC, and interstitial nucleus and its descending fiber extends beyond the TTDC, but no collaterals, arbors, or boutons were found in C1.

ganglion was mainly supported by the indirect data obtained by means of collision (Bahr et al., '81; Pierau et al., '82), double labeling (Taylor and Pierau, '82), or comparisons between the number of cell bodies and axons (Langford and Coggeshall, '81; Suh et al., '84). To our knowledge, the case of peripheral axonal branches visualized by injection of HRP into an individual A-δ neuron was provided only by Hoheisel and Mense ('87). They, however, could find only one neuron having the branches and did not identify its modality. Although our sample was too small to suggest a reliable ratio for dichotomizing peripheral axons, it is certain that a substantial number of A-δ mechanical nociceptive neurons have two or more peripheral axons within the ganglion. We found no such dichotomizing peripheral axons for temperature (Terashima and Liang, '91), touch, or vibrotactile neurons (unpublished data). So dichotomizing peripheral axons seem to be peculiar to nociceptive neurons in this species. In addition, for the neuron with a thinly myelinated and an unmyelinated peripheral axon within the ganglion (Fig. 5), only one AP (conducting in the A-δ range) could be evoked at a fixed latency even if an intense electrical stimulation strong enough to activate C fibers was applied to the peripheral nerve branch, and only one RF could be located in the skin. This indicated that the unmyelinated peripheral axons might be traveling in other nerve branch and supplying deep or visceral structures. Taking all these factors into account, it is reasonable to suggest that dichotomizing peripheral axons of the sensory neurons may be one of the mechanisms of referred pain.

The trigeminal sensory complex in orofacial nociception

It has been suggested that the rostral part of the TSC may not be involved in the sensory aspect of pain, but in motor activities related to pain (Hu and Sessle, '84) or in its motivational affective mechanisms (Melzack and Casey, '68). This view was supported by numerous animal experiments. For example, after tractotomy at the obex level, nociceptive reflex or behavioural reactions induced by stimulation of oral mucosa or facial skin were maintained (Sumino, '71; Young and Perryman, '84; Broton and Rosenfeld, '85) and lesions in the rostral part of the TSC (Rosenfeld et al., '78; Young and Perryman, '84; Pickoff-Matuk et al., '86) or section of the TTDO efferent pathways (Broton and Rosenfeld, '82, '86) resulted in a significant decrease of behavioural reactions to noxious facial stimulation. Although the rostral part of the TSC appears to predominate in orofacial nociceptive reflex or behavioural reactions, its involvement in the nociceptive perception aroused by stimulation of oral mucosa or facial skin can not be ruled out. Electrophysiological studies have shown the existence of nociceptive neurons in the TPR, TTDO, and

TTDI which responded to both nociceptive stimulation of the orofacial region and antidromic electrical stimulation of the contralateral ventrobasal complex of the thalamus, which has connections with the somatosensory cerebral cortex (Eisenman et al., '64; Sessle and Greenwood, '76; Azerad et al., '82). A recent study reported that some neurons in the ventrobasal complex of the thalamus which responded to nociceptive stimulation of oral mucosa or facial skin presented an unchanged response to the identical stimulation after a tractotomy at the obex level (Dallel et al., '88). Also, neurosurgical tractotomy at the same level showed partial persistence or only hypoalgesia of painful oral and facial sensation (McKenzie, '55; Young, '82). In keeping with these physiological and clinical observations, early anatomical studies have shown the direct ascending projections from the rostral part of the TSC to the contralateral and ipsilateral ventrobasal complex of the thalamus in the monkey (Smith, '75), cat (Torvik, '57; Mizuno, '70), and rat (Fukushima and Kerr, '79), and the present study revealed direct nociceptive projections from the orofacial receptive area to the TPR, TTDO, and TTDI. Therefore, it is clear that the rostral part of the TSC is also a relay site which receives primary nociceptive information from oral mucosa and facial skin and sends them directly to both sides of the thalamus to induce a painful sensation.

With respect to the role of the caudal part of the TSC in pain, the TTDC has generally been considered to be of particular importance. On the one hand, the direct ascending pathways from the caudal part of the TSC to the ventrobasal complex of the thalamus, the existence of the TTDC nociceptive neurons responsive to orofacial stimulation and of the TTDC dependent thalamic nociceptive neurons, and the morphological similarities of the TTDC to the spinal dorsal horn have been well documented (for reviews, see Dubner and Bennett, '83; Yokota, '85; Sessle, '89). These qualitative observations provided definite evidence of the involvement of the caudal part of the TSC in nociception. On the other hand, clinical observations, in which trigeminal tractotomy at the obex level often produced complete analgesia (Sjoqvist, '38; McKenzie, '55), have suggested that the TTDC is the essential component of the trigeminal sensory complex concerned with painful orofacial sensation. We found that all the labeled A-δ primary nociceptive neurons projected to the TTDC and the caudal half of the TTDI, but the SC-type neurons did not project to the rostral part of the TSC. These quantitative data demonstrate that the caudal part of the TSC is the major first relay site of orofacial nociception mediated by A-δ neurons.

Another interesting finding in this study is that all the sampled nociceptive neurons innervating oral mucosa and facial skin terminated in the caudal half of the TTDI together with the TTDC and none projected to the TTDC or TTDI only. So the caudal half of the TTDI appears to be of equal importance with the TTDC in orofacial nociception. Marfurt and Del Toro ('87) have reported that transganglionic transport of HRP from the rat central cornea showed a heavy termination in the caudal TTDI and rostral TTDC. Observation of the internuclear pathways of the TSC in the cat has shown that the caudal TTDI and the TTDC had similar input and output connections. For example, the ascending fibers arise only from neurons located in laminae III–V of TTDC and in the caudal TTDI, and the descending fibers do not project to the caudal TTDI or laminae I–II of the TTDC (Nasution and Shigenaga, '87). Our data, however, indicated that trigeminal tractotomy made at the level of the middle of the TTDI and at the level of the obex will result in a different effect on the abolishment of pain sensation, which provides a reasonable explanation of the fact that neurosurgical tractotomy made 8–10 mm rostral to the obex (Sjoqvist, '38) obtained a much better effect in the treatment of trigeminal neuralgia than one made at or caudal to the obex level (McKenzie, '55).

Involvement of the interstitial nucleus in trigeminal nociception

Since Cajal ('09) reported the existence of neurons embedded within the trigeminal descending tract, much study has been devoted to the physiology and morphology of the interstitial nucleus. Because the projection of the trigeminal nerve to the interstitial nucleus has been well documented (Shigenaga et al., '86a,b; Marfurt and Del Toro, '87; Takemura et al., '87; Pfaller and Arvidsson, '88; Marfurt and Rajchert, '91), this nucleus has been considered as a component concerned with trigeminal sensation. However, its precise role in the trigeminal sensory processing remains unclear.

Recent investigations have suggested the involvement of the interstitial nucleus in the sensory processing of orofacial nociceptive information (for a review, see Phelan and Falls, '89). Torvik ('56) reported that the interstitial nucleus is a rostral extension of the superficial laminae of the medullary dorsal horn. Later studies demonstrated that a large region of the interstitial nucleus is cytologically and myeloarchitecturally similar to laminae I and/or II (Hockfield and Gobel, '78; Ishidori et al., '86; Shigenaga et al., '86a,b). Phelan and Falls ('89) reported a similarity in metabolic activity between laminae I and II and the interstitial nucleus on the basis of the cytochrome oxidase pattern of reactivity. Immunocytochemical studies showed that the interstitial nucleus receives prominent input from SP-immunoreactive (Chan-Palay, '78a,b; Sakanaga et al., '82; Shults et al., '84) and enkephalinergic fibers (Fallon and Leslie, '86; Murakami et al., '87). The presence of SP-immunoreactive and enkephalinergic neurons in the interstitial nucleus has also been reported by Chan-Palay ('78a), Finley et al. ('81), and Murakami et al. ('87). By means of the transganglionic transport of HRP, Marfurt and Turner ('84) showed the projections of tooth pulp afferent fibers to the dorsal region of the interstitial nucleus and Shigenaga et al. ('86a) and Marfurt and Del Toro ('87) showed heavy projections of corneal afferent fibers to the interstitial nucleus. Using intra-axonal injection of HRP in the spinal trigeminal tract at the level of the TTDI, Jacquin et al. ('88) observed the terminations of a primary trigeminal A-δ HTM in the interstitial nucleus. Recently, Hayashi and Tabata ('89) reported that the neurons recorded electrophysiologicaly from the interstitial nucleus were all nociceptive specific. The present study revealed that the primary trigeminal A-δ mechanical nociceptive neurons innervating the orofacial region terminated in the interstitial nucleus. Moreover, we have found a total lack of the projections of the temperature neurons (Terashima and Liang, '91) and of the touch, vibrotactile, and pressure neurons (unpublished results) to the interstitial nucleus in the snake.

Therefore, in spite of a few investigations suggesting the involvement of the interstitial nucleus in the processing of orofacial mechanoreceptive, chemoreceptive, and thermoreceptive information (for references, see Phelan and Falls, '89), the studies mentioned above provided abundant evidence to support the assumption that nociceptive information is the major input from the trigeminal nerve to the interstitial nucleus. In view of the projections of the interstitial nucleus to the thalamus and to the parabrachial, solitary, and reticular formation nuclei (see Phelan and Falls, '89, for comprehensive review), it then appears that this nucleus contributes to trigeminal pain perception as well as to motivational and emotional responses to noxious orofacial stimulation.

LITERATURE CITED

Adams, J.C. (1981) Heavy metal intensification of DAB-based HRP reaction product (letter). J. Histochem. Cytochem. 29:775.

Arvidsson, J., and S. Gobel (1981) An HRP study of the central projections of primary trigeminal neurons which innervate tooth pulp in the cat. Brain Res. 208:1–16.

Azerad, J., A. Woda, and D. Albe-Fessard (1982) Physiological properties of neurons in different parts of the cat trigeminal sensory complex. Brain Res. 246:7–11.

Bahr, R., H. Blumberg, and W. Janig (1981) Do dichotomizing afferent fibres exist which supply visceral organs as well as somatic structures? A contribution to the problem of referred pain. Neurosci. Lett. 24:25–28.

Belmonte, C., and R. Gallego (1983) Membrane properties of cat sensory neurons with chemoreceptor and baroreceptor endings. J. Physiol. (Lond.) 342:603–614.

Boyd, I.A., and K.U. Kalu (1979) Scaling factor relating conduction velocity and diameter for myelinated afferent nerve fibers in the cat hindlimb. J. Physiol. (Lond.) 289:277–297.

Broton, J.G., and J.P. Rosenfeld (1982) Rostral trigeminal projections signal perioral facial pain. Brain Res. 243:395–400.

Broton, J.G., and J.P. Rosenfeld (1985) Effects of trigeminal tractotomy on facial thermal nociception in the rat. Brain Res. 333:63–72.

Broton, J.G., and J.P. Rosenfeld (1986) Cutting rostral trigeminal nuclear complex projections preferentially affects perioral nociception in the cat. Brain Res 397:1–8.

Burgess, P.R., and E.R. Perl (1967) Myelinated afferent fibers responding specifically to noxious stimulation of the skin. J. Physiol. (Lond.) 190:541–562.

Burgess, P.R., and E.R. Perl (1973) Cutaneous mechanoreceptors and nociceptors. In A. Iggo (ed): Handbook of Sensory Physiology. Somatosensory System. Heidelberg: Springer, 2:29–78.

Cajal, S. Ramón y (1909) Histologie du systeme nerveux de l'homme et des vertebres. Vol. I. Paris: Maloine.

Chan-Palay, V. (1978a) The paratrigeminal nucleus: I. Neurons and synaptic organization. J. Neurocytol. 7:405–418.

Chan-Palay, V. (1978b) The paratrigeminal nucleus: II. Identification and inter-relations of catecholamine axons, indoleamine axons, and substance P immunoreactive cells in the neuropil. J. Neurocytol. 7:419–442.

Cooper, B., M. Ahlquist, R.M. Friedman, B. Loughner, and M. Heft (1991) Properties of high-threshold mechanoreceptors in the oral mucosa. I. Responses to dynamic and static pressure. J. Neurophysiol. 66:1272–1279.

Dallel, R., P. Raboisson, P. Auroy, and A. Woda (1988) The rostral part of the trigeminal sensory complex is involved in orofacial nociception. Brain Res. 448:7–19.

Dallel, R., P. Raboisson, A. Woda, and B.J. Sessle (1990) Properties of nociceptive and non-nociceptive neurons in trigeminal subnucleus oralis of the rat. Brain Res. 521:95–106.

Dubner, R., S. Gobel, and D.D. Price (1976) Peripheral and central trigeminal pain pathways. In J.J. Bonica and D. Albe-Fessard (eds): Advances in Pain Research and Therapy. New York: Raven Press, 1:137–148.

Dubner, R., and J.W. Hu (1977) Myelinated (Aδ) nociceptive afferents innervating the monkey's face. J. Dent. Res. 56:A167.

Dubner, R., and G.J. Bennett (1983) Spinal and trigeminal mechanisms of nociception. Annu. Rev. Neurosci. 6:381–418.

Eisenman, J., G. From, J. Landgren, and D. Novin (1964) The ascending projections of trigeminal neurons in the cat investigated by antidromic stimulation. Acta Physiol. Scand. 60:337–350.

Fallon, J.H., and F.M. Leslie (1986) Distribution of dynorphin and enkephalin peptides in the rat brain. J. Comp. Neurol. 249:293–336.

Finley, J.C.W., J.L. Maderdrut, L.J. Roger, and P. Petrusz (1981) The
immunocytochemical localization of somatostatin-containing neurons in the rat central nervous system. Neuroscience 7:2173–2192.

Fukushima, T., and F.W.L. Kerr (1979) Organization of trigemino-thalamic tracts and other thalamic afferent systems of the brainstem in the rat. Presence of gelatinosa neurons with thalamic connections. J. Comp. Neurol. 183:169–184.

Gasser, H.S. (1955) Properties of dorsal root unmedullated fibers on both the sides of the ganglion. J. Gen. Physiol. 38:709–728.

Gorke, K., and F.-K. Pierau (1980) Spike potentials and membrane properties of dorsal root ganglion cells in pigeons. Pflugers Arch. 386:21–28.

Grant, F.C. (1955) Discussion on trigeminal tractotomy. Clin. Neurosurg. 2:69–70.

Gurtu, S.G., and P.A. Smith (1988) Electrophysiological characteristics of hamster dorsal root ganglion cell and their response to axotomy. J. Neurophysiol. 59:408–423.

Handwerker, H.O., F. Anton, and P.W. Reeh (1987) Discharge patterns of afferent cutaneous nerve fibers from the rat's tail during prolonged noxious mechanical stimulation. Exp. Brain Res. 65:493–504.

Harper, A.A., and S.N. Lawson (1985) Electrical properties of rat dorsal root ganglion neurons with different peripheral nerve conduction velocities. J. Physiol. (Lond.) 359:47–63.

Hayashi, H., R. Sumino, and B.J. Sessle (1984) Functional organization of trigeminal subnucleus interpolaris: nociceptive and innocuous afferent inputs, projections to thalamus, cerebellum and spinal cord, and descending modulation from periaqueductal gray. J. Neurophysiol. 51:890–905.

Hayashi, H., and T. Tabata (1989) Physiological properties of sensory neurons of the interstitial nucleus in the spinal trigeminal tract. Exp. Neurol. 105:219–220.

Hockfield, S., and S. Gobel (1978) Neurons in and near nucleus caudalis with long ascending projection axons demonstrated by retrograde labeling with horseradish peroxidase. Brain Res. 139:333–339.

Hoheisel, U., and S. Mense (1987) Observations on the morphology of axons and somata of slowly conducting dorsal root ganglion cells in the cat. Brain Res. 423:269–278.

Hu, J.W., and B.J. Sessle (1984) Comparison of responses of cutaneous nociceptive and non-nociceptive brain stem neurons in trigeminal subnucleus caudalis (medullary dorsal horn) and subnucleus oralis to natural and electrical stimulation of tooth pulp. J. Neurophysiol. 52:39–53.

Hu, J.W., and B.J. Sessle (1988) Properties of functionally identified nociceptive and nonnociceptive facial primary afferents and presynaptic excitability changes induced in their brain stem endings by raphe and orofacial stimuli in cats. Exp. Neurol. 101:385–399.

Hursh, J.B. (1939) Conduction velocity and diameter of nerve fibers. Am. J. Physiol. 127:131–139.

Ishidori, H., T. Nishimori, Y. Shigenaga, S. Suemune, Y. Dateoka, M. Sera, and N. Nagasaka (1986) Representation of upper and lower primary teeth in the trigeminal sensory nuclear complex in the young dog. Brain Res. 370:153–158.

Jacquin, M.F., R.A. Stennett, W.E. Renehan, and R.W. Rhoades (1988) Structure-function relationships in the rat brainstem subnucleus interpolaris. II. Low and high threshold trigeminal primary afferents. J. Comp. Neurol. 267:107–130.

Jacquin, M.F., N.L. Chiaia, J.H. Haring, and R.W. Rhoades (1990) Intersubnuclear connections within the rat trigeminal brainstem complex. Somatosens. Mot. Res. 7:399–420.

Khayyat, G.F., Y.J. Yu, and R.B. King (1975) Response patterns to noxious and non-noxious stimuli in rostral trigeminal relay nuclei. Brain Res. 97:47–60.

Koerber, H.R., R.E. Druzinsky, and L.M. Mendell (1988) Properties of somata of spinal dorsal root ganglion cells differ according to peripheral receptor innervated. J. Neurophysiol. 60:1584–1596.

Kruger, L., E.R. Perl, and M.J. Sedives (1981) Fine structure of myelinated and mechanical nociceptor endings in cat hairy skin. J. Comp. Neurol. 198:137–154.

Langford, L.A., and R.E. Coggeshall (1981) Branching of sensory axons in the peripheral nerve of the rat. J. Comp. Neurol. 203:745–750.

Lee, K.H., K. Chung, J.M. Chung, and R.E. Coggeshall (1986) Correlation of cell body size, axon size, and signal conduction velocity for individually labeled dorsal root ganglion cells in the cat. J. Comp. Neurol. 234:335–346.

Lieberman, A.R. (1976) Sensory ganglion in the peripheral nerve. In D.N. Landon (ed): The Peripheral Nerve. London: Chapman and Hall, pp. 188–278.

Loeb, G.E. (1976) Decreased conduction velocity in the proximal projections of myelinated dorsal root ganglion cells in cat. Brain Res. 103:381–385.

Lynn, B., and S.E. Carpenter (1982) Primary afferent units from the hairy skin of the rat hind limb. Brain Res. 238:29–34.

Maciewicz, R., P. Mason, A. Strassman, and S. Potrebic (1988) Organization of trigeminal nociceptive pathways. Semin. Neurol. 8:255–264.

Marfurt, C.F., and D.F. Turner (1984) The central projections of tooth pulp afferent neurons in the rat as determined by the transganglionic transport of horseradish peroxidase. J. Comp. Neurol. 223:535–547.

Marfurt, C.F., and D.R. Del Toro (1987) Corneal sensory pathways in the rat: A horseradish peroxidase tracing study. J. Comp. Neurol. 261:450–459.

Marfurt, C.F., and D.M. Rajchert (1991) Trigeminal primary afferent projections to "non-trigeminal" areas of the rat central nervous system. J. Comp. Neurol. 303:489–511.

Martin, H.A., A.I. Basbaum, G.C. Kwiat, E.J. Goetzl, and J.D. Levine (1987) Leukotriene and prostaglandin sensitization of cutaneous high-threshold C- and A-delta mechanonociceptors in the hairy skin of rat hindlimbs. Neuroscience 22:651–659.

McGrath, P.A., R. Gracely, R. Dubner, and M. Heft (1983) Non-pain and pain sensations evoked by tooth pulp stimulation. Pain 15:377–388.

McKenzie, K.G. (1955) Trigeminal tractotomy. Clin. Neurosurg. 2:50–69.

Melzack, R., and K.L. Casey (1968) Sensory, motivational and central control determinants of pain. In D.R. Kenshalo (ed): The Skin Senses. Springfield, IL: Thomas, pp. 423–443.

Mizuno, N. (1970) Projection fibers from the main sensory trigeminal nucleus and the supratrigeminal region. J. Comp. Neurol. 139:457–471.

Molenaar, G.J. (1978a) The sensory trigeminal system of a snake in the possession of infrared receptors. I. The sensory trigeminal nuclei. J. Comp. Neurol. 179:123–136.

Molenaar, G.J. (1978b) The sensory trigeminal system of a snake in the possession of infrared receptors. II. The central projections of the trigeminal nerve. J. Comp. Neurol. 179:137–152.

Murakami, S., M. Okamura, C. Yanaihara, N. Yanaihara, and Y. Ibata (1987) Immunocytochemical distribution of met-enkephalin-Arg-Gly-Leu in the rat lower brainstem. J. Comp. Neurol. 261:193–208.

Nasution, I.D., and Y. Shigenaga (1987) Ascending and descending internuclear projections within the trigeminal sensory nuclear complex. Brain Res. 425:234–247.

Perl, E.R. (1968) Myelinated afferent fibres innervating the primate skin and their response to noxious stimuli. J. Physiol. (Lond.) 197:593–615.

Pfaller, K., and J. Arvidsson (1988) Central distribution of trigeminal and upper cervical primary afferents in the rat studied by anterograde transport of horseradish peroxidase conjugated to wheat germ agglutinin. J. Comp. Neurol. 268:91–108.

Phelan, K.D., and W.M. Falls (1989) The interstitial system of the spinal trigeminal tract in the rat: Anatomical evidence for morphological and functional heterogeneity. Somatosens. Mot. Res. 6:367–399.

Pickoff-Matuk, J.F., J.P. Rosenfeld, and J.G. Broton (1986) Lesions of the mid-spinal trigeminal complex are effective in producing perioral thermal hypoalgesia. Brain Res. 382:291–298.

Pierau, Fr.-K., D.C.M. Taylor, W. Abel, and B. Friedrich (1982) Dichotomizing peripheral fibers revealed by intracellular recording from rat sensory neurons. Neurosci. Lett. 31:123–128.

Reeh, P.W., J. Bayer, L. Kocher, and H.O. Handwerker (1987) Sensitization of nociceptive cutaneous nerve fibers from the rat's tail by noxious mechanical stimulation. Exp. Brain Res. 65:505–512.

Rose, R.D., H.R. Koerber, M.J. Sedivec, and L.M. Mendell (1986) Somal action potential duration differs in identified primary afferents. Neurosci. Lett. 63:259–264.

Rosenfeld, J.P., R.M. Clavier, and J.G. Broton (1978) Bilateral and unilateral antinociceptive effects of rostral trigeminal nuclear complex lesions in rats. Brain Res. 157:147–152.

Sakanaga, M., S. Inagaki, S. Shiosaka, E. Senba, H. Takagi, K. Takatsuki, Y. Kawai, H. Iida, Y. Hara, and M. Tohyama (1982) Ontogeny of substance P-containing neuron systems of the rat: Immunohistochemical analysis. II. Lower brain stem. Neuroscience 7:1097–1126.

Sessle, B.J., and L.F. Greenwood (1976) Inputs to trigeminal brain stem neurons from facial, oral, tooth pulp and pharyngolaryngeal tissues. I. Responses to innocuous and noxious stimuli. Brain Res. 117:211–226.

Sessle, B.J. (1989) Neural mechanisms of oral and facial pain. Otolaryngol. Clin. North Am. 22:1059–1072.

Shigenaga, Y., I.C. Chen, S. Suemune, T. Nishimori, I.D. Nasution, A. Yoshida, H. Sato, T. Okamoto, M. Sera, and M. Hosoi (1986a) Oral and facial representation within the medullary and upper cervical dorsal horn in the cat. J. Comp. Neurol. 243:388–408.

Shigenaga, Y., T. Okamoto, T. Nishimori, S. Suemune, I.D. Nasution, I.C. Chen, K. Tsuru, A. Yoshida, K. Tabushi, M. Hosoi, and H. Tsuru (1986b) Oral and facial representation in the trigeminal principal and rostral spinal nuclei of the cat. J. Comp. Neurol. 244:1–18.

Shults, C.W., R. Quirion, B. Chronwall, T.N. Chase, and T.L. O'Donohue (1984) A comparison of the anatomical distribution of substance P and substance P receptors in the rat central nervous system. Peptides 5:1097–1128.

Sinclair, D.C., G. Weddell, and H. Feindel (1948) Referred pain and associated phenomena. Brain 71:184–211.

Sjoqvist, O. (1938) Studies on pain conduction in the trigeminal nerve. A contribution to surgical treatment of facial pain. Acta Psychiatr. Neurol. Scand. [Suppl.] 17:1–139.

Smith, R.L. (1975) Axonal projection and connections of the principal sensory trigeminal nucleus in the monkey. J. Comp. Neurol. 163:347–376.

Spencer, P.S., C.S. Raine, and H. Wisniewski (1973) Axon diameter and myelin thickness—unusual relationships in dorsal root ganglia. Anat. Rec. 176:225–244.

Stewart, W.A., and R.B. King (1963) Fiber projections from the nucleus caudalis of the spinal trigeminal nucleus. J. Comp. Neurol. 121:271–286.

Stoney, S.D., Jr. (1985) Unequal branch point filtering action in different types of dorsal root ganglion neurons of frogs. Neurosci. Lett. 59:15–20.

Suh, Y.S., K. Chung, and R.E. Coggeshall (1984) A study of axonal diameters and areas in lumbosacral roots and nerves in the rat. J. Comp. Neurol. 222:473–481.

Sumino, R. (1971) Central neural pathways involved in the jaw-opening reflex in the cat. In R. Dubner and Y. Kawamura (eds): Oral Facial Sensory and Motor Mechanisms. New York: Appleton-Century-Crofts, pp. 315–331.

Takemura, M., T. Sugimoto, and A. Sakai (1987) Topographic organization of central terminal region of different sensory branches of the rat mandibular nerve. Exp. Neurol. 96:540–557.

Taylor, D.C.M., and Fr.-K. Pierau (1982) Double fluorescence labeling supports electrophysiological evidence for dichotomizing peripheral sensory nerve fibers in rat. Neurosci. Lett. 33:1–6.

Terashima, S., and Y.-F. Liang (1991) Temperature neurons in the crotaline trigeminal ganglia. J. Neurophysiol. 66:623–634.

Terashima, S., and Y.-F. Liang (1992) Central projections of Aδ nociceptive neurons in the trigeminal ganglion of the crotaline snake. J. Physiol. (Lond.) 446:158P.

Torvik, A. (1956) Afferent connections to the sensory trigeminal nuclei, the nucleus of the solitary tract and adjacent structures. J. Comp. Neurol. 106:51–141.

Torvik, A. (1957) Ascending fibers from the main trigeminal sensory nucleus. An experimental study in the cat. Am. J. Anat. 100:1–16.

Traub, R.J., and L.M. Mendell (1988) The spinal projection of individual identified A-δ- and C-fibers. J. Neurophysiol. 59:41–55.

Yokota, T. (1985) Neural mechanisms of trigeminal pain. In H.L. Fields, R. Dubner, and F. Cervero (eds): Advances in Pain Research and Therapy. New York: Raven Press, 9:211–232.

Young, R.F. (1982) Effects of trigeminal tractotomy on dental sensation in humans. J. Neurosurg. 56:812–818.

Young, R.F., and K.M. Perryman (1984) Pathways for orofacial pain sensation in the trigeminal brain-stem nuclear complex of the macaque monkey. J. Neurosurg. 61:563–568.

Young, R.F., and K.M. Perryman (1986) Neuronal responses in rostral trigeminal brain-stem nuclei of macaque monkeys after chronic trigeminal tractotomy. J. Neurosurg. 65:508–516.

Visual and infrared input to the same dendrite in the tectum opticum of the python, *Python regius*: electron-microscopic evidence

Sonou Kobayashi [a], Reiji Kishida [b], Richard C. Goris [a], Masami Yoshimoto [c] and Hironobu Ito [c]

[a] *Department of Anatomy, School of Medicine, Yokohama City University, Yokohama (Japan), [b] Department of Anatomy, School of Medicine, Yamaguchi University, Ube (Japan) and [c] Department of Anatomy, Nippon Medical School, Tokyo (Japan)*

Key words: Python; Optic tectum; Infrared–visual correlation; Bimodal neuron; Horseradish peroxidase; Degeneration

In snakes with infrared receptors, the optic tectum receives input from both the visual and the infrared senses. We investigated the infrared and optic fiber terminations in the tectum with a combination of horseradish peroxidase and degeneration labeling. In addition to synapses by visual and infrared fibers onto individual neurons, we were able to observe for the first time visual and infrared synapses on one and the same dendrite.

The question of how various types of sensory input influence each other in the central nervous system and how the type of reaction is determined is as yet unsolved. The optic tectum receives several types of sensory input and is one of the largest correlational centers in the central nervous system. However, it is not yet known how the optic tectum processes this information.

In snakes with infrared receptors the optic tectum receives both infrared and visual input[4,8,13]. However, there are no reports on synaptic correlation between retinal and infrared input or their receptive neurons in the optic tectum. We used pythons to investigate how these two different types of information relate to each other in the optic tectum.

In pit vipers, input from infrared receptor cells in the pit organ passes through the trigeminal ganglion to the lateral nucleus of the medulla oblongata, i.e., the nucleus of the lateral descending trigeminal tract (LTTD)[5,13], and to the reticularis caloris[2], and then terminates in the contralateral optic tectum[2,7,10]. In pythons and boas this route runs from the trigeminal ganglion through the LTTD to the contralateral optic

tectum[5,7]. Input from the eyes passes through the optic nerve to the contralateral optic tectum[1,8].

In the present study we used horseradish peroxidase (HRP) labeling and nerve degeneration to observe simultaneously in the same individual the optic and infrared fiber connections in the tectum opticum.

Three pythons, *Python regius*, about 100 cm in length and 1 kg in weight were used. Under general anesthesia with a modified flow-through method[9], we ablated the temporal muscles from the skull and retracted them. Next we severed the quadrate from the skull and opened a hole with a dental engine and a round burr through the inner ear to expose the medulla. The meninges were slit with a hypodermic needle. Thirty percent HRP (Toyobo Co., Japan) in 1 μl of distilled water containing 2.5% L-α-lysophosphatidylcholine (Sigma) was slowly (over a period of 30 min) injected into the left LTTD using a microsyringe on a manipulator. Next an incision was made in the muscles around the left eyeball and the optic canal was exposed. The optic nerve with the central artery and vein was freed from the surrounding tissues, and then ligated and cut with the blood vessels.

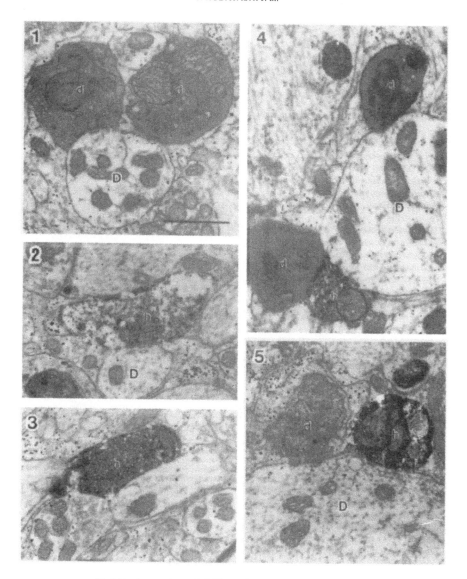

Fig. 1. Two degenerating terminals (d) synapsing with a dendrite (D). Scale = 1 μm.

Fig. 2. An HRP-labeled terminal (h) synapsing with a dendrite (D).

Fig. 3. An HRP-labeled terminal (h) making a contact with another terminal of unknown origin.

Fig. 4. Two degenerating terminals (d) and an HRP-labeled terminal (h) synapsing with a dendrite (D).

Fig. 5. Degenerating (d) and HRP-labeled (h) terminals synapsing with a dendrite (D).

After the operation each python was given two intramuscular injections of 4 mg/kg of amikacin sulfate at an interval of 2 days and kept alive for 7 days at a temperature of 25°C and 50% humidity with free access to drinking water. Then the animal was anesthetized with fluothane and perfused through the heart with heparinized (5 U/ml) 0.9% saline, followed by perfusion with 2% paraformaldehyde and 2% glutaraldehyde in 0.2 M phosphate buffer at pH 7.4. The brains were then removed, and frontal serial slices of the right optic tectum were cut at 200 μm, processed for HRP with DAB, and postfixed in 2% osmic acid. After dehydration with ethanol, the slices were embedded in a mixture of Epon and Araldite.

Degenerating terminals from the retina formed synapses on dendrites of various sizes in layers 7a–13 (Fig. 1). HRP-labeled terminals from the LTTD made synaptic contacts with dendrites in layers 5–13 (Fig. 2). Some HRP-labeled terminals made synaptic contact with other terminals of unknown origin (Fig. 3). Most of these terminals were found on separate dendrites. In layers 7b and 8, however, both degenerating and HRP-labeled terminals formed synapses on a single dendrite. These two synaptic sites were in close proximity (Figs. 4, 5).

Using HRP, Schroeder[11] reported that retinal fibers terminate in the stratum zonale (SZ), the stratum opticum (SO), and the stratum griseum et fibrosum centrale (SFGC) in the rattlesnake (Crotalus viridis). Molenaar[6] used degeneration techniques to demonstrate the projection of infrared sensory neurons to the tectum opticum in pythons (Python reticulatus). Newman et al.[7] used HRP to demonstrate such projections in both pythons and pit vipers (Crotalus). Kishida et al.[10] likewise demonstrated infrared tectal projections in oriental pit vipers (Agkistrodon). With electrophysiological recording and marking techniques, Kass et al.[4] were able to demonstrate the existence of, and localize, three types of neurons in the tectum: visual, infrared, and bimodal visual–infrared. However, to date there has been no direct histological evidence of proven visual or infrared fibers synapsing with tectal neurons. In particular, it was unknown whether the bimodal neurons received direct input from extratectal neurons, or received their input from interneurons within the tectum. The present work demonstrates for the first time direct synapsing by extratectal neurons onto tectal dendrites (Figs. 1–3). In particular we present evidence here of both visual and infrared extratectal fibers synapsing in close proximity to each other onto one and the same dendrite (Figs. 4, 5).

Of course, this does not exclude the possibility that interneurons also exist. But the fact that we were able to demonstrate such synapsing on bimodal neurons in both cross (Fig. 4) and longitudinal (Fig. 5) sections in the same material is evidence that this type of synapsing occurs there with a high degree of frequency.

A part of this research was supported by Grant 01440019 from the Ministry of Education of Japan to H.I. for scientific research.

1 Armstrong, J.A., An experimental study of the visual pathways in a snake (Natrix natrix), J. Anat., 85 (1951) 275–288.
2 Gruber, E.R., Kicliter, E., Newman, E.A., Kass, L. and Hartline, P.H., Connections of the tectum of the rattlesnake Crotalus viridis: an HRP study, J. Comp. Neurol., 188 (1979) 31–42.
3 Hartline, P.H., Kass, L. and Loop, M.S., Merging of modalities in the optic tectum: infrared and visual integration in rattlesnakes, Science, 199 (1978) 1225–1229.
4 Kass, L., Loop, M.S. and Hartline, P.H., Anatomical and physiological localization of visual and infrared cell layers in tectum of pit vipers, J. Comp. Neurol., 182 (1978) 811–820.
5 Molenaar, G.J., An additional trigeminal system in certain snakes possessing infrared receptors, Brain Res., 78 (1974) 340–344.
6 Molenaar, G.J. and Fizaan-Oostveen, J.L.F.P., Ascending projections from the lateral descending and common sensory trigeminal nuclei in python, J. Comp. Neurol., 189 (1980) 555–572.
7 Newman, E.A., Gruber, E.R. and Hartline, P.H., The infrared trigemino-tectal pathway in the rattlesnake and in the python, J. Comp. Neurol., 191 (1980) 465–477.
8 Northcutt, R.G., Anatomical organization of the optic tectum in reptiles. In H. Vanegas (Ed.), Comparative Neurology of the Optic Tectum, Plenum Press, New York/London, 1984, pp. 548–600.
9 Goris, R.C. and Terashima, S., Central response to infrared stimulation of the pit receptors in a crotaline snake, Trimeresurus flavoviridis, J. Exp. Biol., 58 (1973) 59–76.
10 Kishida, R., Amemiya, F., Kusunoki, T. and Terashima, S., A new tectal afferent nucleus of the infrared sensory system in the medulla oblongata of Crotaline snakes, Brain Res., 195 (1980) 271–279.
11 Schroeder, D.M., Retinal afferents and infrared sensitive snake, Crotalus viridis, J. Comp. Neurol., 170 (1981) 29–42.
12 Schroeder, D.M. and Loop, M.S., Trigeminal projections in snakes possessing infrared sensitivity, J. Comp. Neurol., 169 (1976) 1–14.
13 Terashima, S. and Goris, R.C., Tectal organization of pit viper infrared reception, Brain Res., 83 (1975) 490–494.

Innervation of Snake Pit Organ Membranes Mapped by Receptor Terminal Succinate Dehydrogenase Activity

Richard C. GORIS, Tetsuo KADOTA and Reiji KISHIDA

Yokohama City University, School of Medicine, Fukuura 3-9, Kanazawa-ku, Yokohama, 236 JAPAN

Abstract: Using the mamushi, *Agkistrodon blomhoffii*, we stained the succinate dehydrogenase activity in the infrared receptors of the pit organs by whole-body perfusion with nitro blue tetrazolium. This permitted light microscope observation of the individual receptors, or terminal nerve masses (TNMs). In whole mounts TNMs measured 20–150 μm in long diameter. The longer TNMs were possibly several receptors seen together, because serial sections showed overlapping of the edges of adjacent receptors. TNMs were also present in the scales at the dorsal and caudal edge of the pit membrane. In degeneration experiments, the perfusion technique permitted mapping of the innervation of the pit membrane by trigeminal branches: the ophthalmic branch innervated the dorsocaudal 18%, the deep branch of the maxillary the dorsorostral 46%, and the superficial branch of the maxillary the ventral 36% of the membrane. We also counted the number of myelinated fibers in these nerve trunks. Although the deep branch innervated the largest portion of the membrane, it had the fewest fibers (815, vs. 1,206 for the ophthalmic and 1,832 for the superficial). This fact, in connection with the homogeneous size of the deep branch fibers, suggested that the deep branch had the purest population of infrared fibers.

Key Words: Pit organs; Infrared receptors; Terminal nerve masses; Succinate dehydrogenase; Nitro blue tetrazolium

The infrared receptors in the pit organs of crotaline snakes are specialized nerve terminals called "terminal nerve masses" (TNMs) located just beneath the outer epithelium of the pit membrane (Terashima et al., 1970). The morphology of these TNMs has been studied with the electron microscope (Terashima et al., 1970; Hirosawa, 1980), but to date there has been no adequate means of staining the individual receptors for observation with the light microscope. These nerve terminals contain heavy concentrations of mitochondria (De Robertis and Bleichmar, 1962; Terashima et al., 1970; Hirosawa, 1980), which in turn contain succinate dehydrogenase (SDH), whose activity can be demonstrated by reaction with nitro blue tetrazolium (Nachlas et al., 1957; Ogawa and Barrnett, 1964; Rosa and Tsou, 1965). However, it is difficult to stain the isolated pit membrane by standard procedures, because the receptor layer is sandwiched between two hydrophobic cornified layers; and removal of either of the layers can result in distortion of the underlying structures. We therefore developed a whole-body perfusion technique to stain the pit organ mitochondria by the SDH reaction, in order to study by light microscopy the distribution of the receptors in the pit membrane and their innervation.

MATERIALS AND METHODS

For the demonstration of SDH we used adult pit vipers called "mamushi", *Agkistrodon blomhoffii*, from the Asian mainland. These animals were first anesthetized with halothane and then perfused through the right aortic arch with 200 cc of 0.9% saline solution containing 5 i.u./cc heparin and 0.02% of an additional anesthetic, tricaine methanesulfonate. The animals were next perfused with 400 cc of 0.05 M phosphate buffer (pH 7.4) containing 0.05 M sodium succinate and 200 mg of nitro blue tetrazolium. The flow rate was controlled to take 30 min for this latter perfusion. Both the saline and the reagent fluids were maintained at 37 C, and the animal's body was kept immersed in a 40 C water bath throughout the reaction. The temperature of the head as measured within the mouth was a constant 37 C. The progress of the reaction in the pit membrane was monitored directly with a dissection microscope. finally, to stop the reaction and fix the tissues, the animal was perfused with 400 cc of 10% formalin.

Next the pit membranes were dissected out intact, together with the scales of the mouth of the pit to which each membrane was attached. The membranes were first photographed in this state under a low-power microscope. Then several incisions were made in some of the membranes so that they would lie flat on glass slides. These specimens were then dehydrated in an ethanol series, cleared with xylene, and mounted whole on glass slides with a coverslip and Canadian balsam. Other membranes were frozen, cross-sectioned serially with a cryostat, and mounted on gelatinized glass slides.

We also combined this technique with degeneration techniques to visualize the area of the pit membrane innervated by each of the three nerves supplying the pit, i.e., the ophthalmic branch and the deep and superficial branches of the maxillary branch of the trigeminal nerve (Kishida et al., 1982). Under halothane and ketamine anesthesia, either the superficial branch alone or the deep and superficial branches together were severed. We did not attempt to cut the ophthalmic branch because of its almost inaccessible location inside the skull and orbit. After a survival time of 7–14 days, the snakes were perfused with reagent fluid and the pit membranes photographed, mounted, or sectioned as described above.

We next prepared camera lucida drawings of the serial cross sections of specimens altered by degeneration, and measured the degenerated area of each membrane with a digitizer connected to a personal computer (Canon CX-1). In this way the relative area innervated by each branch could be calculated. finally, to correlate the measured areas with the size of the nerve supply, we also prepared cross sections of each of the trigeminal branches supplying the pit. After perfusing the snakes with glutaraldehyde fixative, we postfixed the nerves with osmium and embedded them in Epon in the conventional procedure for electron microscope viewing. We cut 5 μm cross sections from a part of the nerves as close to the pit as possible before

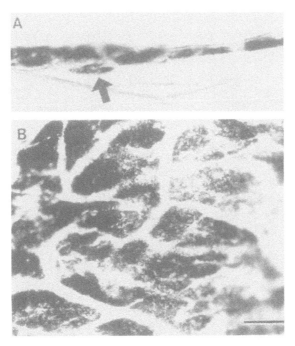

Fig. 1. Infrared receptor terminal nerve masses (TNMs) visualized by the succinate dehydroge-nase activity of their mitochondria. A: cross section through a pit membrane. Stained TNMs are arranged in a single layer, but overlapping of adjacent receptors is apparent, especially at the arrow. B: Whole mount of a pit membrane showing stained TNMs of various shapes and sizes bordered by capillaries (white spaces). A and B: × 320. Bar = 50 μm.

extensive branching began, and stained them with toluidine blue. We then made camera lucida drawings under the light microscope and counted the number of myelinated nerve fibers in each.

RESULTS

With the techniques described, the terminal nerve masses (TNMs) in the pit membrane were stained a deep purple (Fig. 1). Capillary blood vessels ap-peared as clear spaces separating the TNMs. Nerve fibers and other structures were not stained at all.

The staining enabled us to view clearly with the light microscope, even at low magnification, the condition in situ of the TNMs in the infrared-sensi-tive membrane of the pit organs. The TNMs were densely distributed throughout the membrane. They were also present in small quantities in the scales at the dorsal and caudal edges of the pit menbrane, but not in the scales at the rostral or ventral edges. Many were separated from each other and de-lineated by blood vessels; but others were often so close together that it was not possible to distinguish whether the object was several small TNMs closely pressed together, or one large TNM with indented borders (Fig. 1).

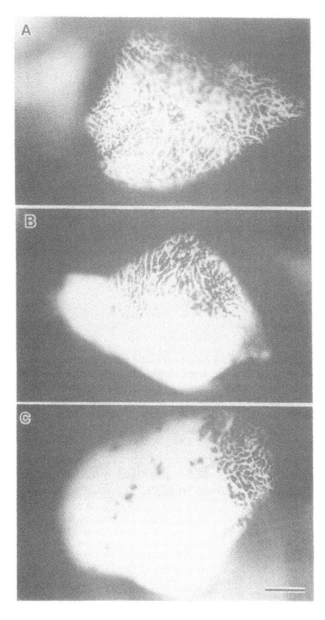

Fig. 2. The effects of degeneration experiments seen in the pit membrane still attached to the scales of the mouth of the pit. A: intact, control membrane. Rostral is to the right of this specimen. B: The superficial branch of the maxillary has been cut. Succinate dehydrogenase activity has disappeared from the ventral part of this membrane, but remains in the areas innervated by the 2 intact branches of the trigeminal nerve. Rostral is to the left. C: Both superficial and deep branches of the maxillary have been cut. Activity remains only in the dorsocaudal part innervated by the ophthalmic branch. The objects inside the white area are not receptor terminals but pigmented spots. Rostral is to the left. A, B, and C: × 32. Bar = 500 μm.

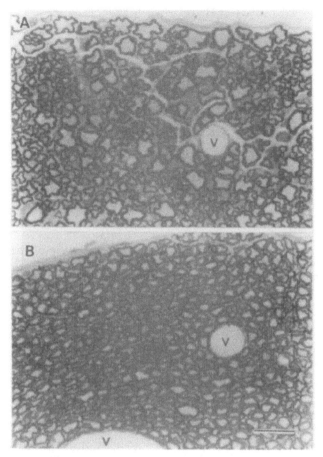

Fig. 3. Cross section through 2 nerve trunks supplying the pit membrane. A: superficial branch of the maxillary. There is a wide range of sizes of myelinated fibers. B: deep branch of the maxillary. Most of the fibers appear to be in the same size class. V, blood vessel. A and B: × 640. Bar = 25 μm.

The reason for this was the overlapping of the edges of individual TNMs. Although the TNMs were arranged in a single layer, the edges of neighboring TNMs frequently overlapped. This could be clearly seen in cross sections through the pit membrane (Fig. 1A). Therefore it was not possible to count the number of individual TNMs. The smallest clearly defined TNMs were 20 μm in diameter, and the largest seemingly delimited ones were 150 μm in long diameter; but these larger measurements are unreliable for the reasons stated.

The results of the degeneration studies were very clear-cut. Each nerve branch innervated a clearly defined area of the pit membrane (Figs. 2 and 4). The superficial branch of the maxillary branch innervated the entire ventral section; the deep branch of the maxillary branch innervated the dorsorostral

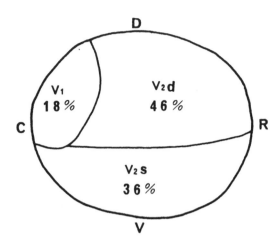

Fig. 4. Diagram showing the location and relative size of areas of the pit membrane innervated by each of its 3 nerve trunks. C, caudal; D, dorsal; R, rostral; V, ventral; V_1, ophthalmic branch; V_2d, deep branch of the maxillary; V_2s, supercial branch of the maxillary.

section; and the ophthalmic branch innervated the dorsocaudal section, which is the most forward-looking portion of the membrane.

The average diameter of the membranes measured was 2.3 mm, and the average area was 4.6 mm² measured along the curve of the membrane in serial sections. The proportion of this area innervated by each nerve branch, as measured from camera lucida tracings of serial cross sections was as follows: ophthalmic branch, 18%; deep branch, 46%; and superficial branch, 36%. In these measurements we included only the stained and unstained parts of the free membrane proper, and excluded those edges of the pit mouth which contained stained TNMs.

The number of myelinated fibers in the trigeminal branches feeding the membrane, as counted in representative cross sections, were as follows: ophthalmic branch, 1,206; deep branch of the maxillary branch, 865; and superficial branch of the maxillary branch, 1,832. This is a total of 3,903 fibers. fiber size was fairly homogeneous in the deep branch, but varied considerably in the ophthalmic and superficial branches (Fig. 3).

DISCUSSION

This is the first time that the TNMs have been adequately stained for viewing with the light microscope. Lynn (1931) and Cordier et al. (1964) studied the histology of the pit with various stains under the light microscope, and described what they believed to be nucleated receptor cells. Since the only nucleated cells in the receptor layer of the membrane are Schwann cells, they were clearly mistaking these cells for receptors (Terashima et al., 1970). Bullock and Fox (1957) also studied the histology of the pit membrane under the light microscope using silver staining. They described structures looking like the palm of a hand with naked nerve fibrils extending from it. Each of their "palmate structures" with its associated fibrils taken as a whole was probably a TNM, but because of the selective nature of their silver stain, which stained mainly the skeletal structure of the nerves, they did not visualize the TNMs as we know them today. Hirosawa (1980), using an Asian crotaline, *Trimeresurus flavoviridis*, was able to duplicate the results of Bullock and Fox (1957), and showed that with silver staining, the TNMs appear as palmate structures with naked nerve fibrils.

The abundant SDH activity revealed in the TNMs by our procedures confirmed previous electron microscope observations that the infrared receptors contain a heavy population of mitochondria (De Robertis and Bleichmar, 1962; Terashima et al., 1970; Hirosawa, 1980). This heavy mitochondrial population seems to be a characteristic of the primary afferent neurons of the infrared sensory system, for Kishida et al. (unpublished observations) found evidence of SDH activity in the trigeminal ganglion of *A. blomhoffii*, and Kusunoki et al. (1987) found strong SDH activity in the same snake in the lateral descending trigeminal nucleus of the medulla, where the primary infrared neurons have their terminals.

The presence of TNMs in the scales at the edge of the pit has not been reported in any previous work. This is probably because previous workers stained pit membranes that had been first dissected out from the pits. However, Terashima and Goris (1979) noted the presence of two receptive areas in a scale at the caudal edge of the pit, so the present discovery is not an altogether unexpected one.

We had originally hoped to be able to count and measure the individual TNMs using the SDH reaction, but we found this to be next to impossible in whole mounts because of the frequent edge overlapping in adjacent receptors. Both Bullock and Fox (1957) and Hirosawa (1980) observed such overlapping between the palmate structures of their silver-stained preparations.

Terashima et al. (1970) reported a maximum diameter of 40 μm for the TNMs. Our observations seemed to indicate that the TNMs are very inhomogeneous both in size and shape. We measured sizes of 20–150 μm in long diameter. However, Terashima et al. (1970) reported that each TNM was separated from its neighbors by connective tissue, and since connective tissue was not stained by our methods, it was not always possible to judge with certainty whether a given TNM was single or actually made up of several slightly overlapping TNMs.

Despite this overlapping, however, our work confirmed the light microscopic observations of Bullock and Fox (1957) and the electron microscopic observations of Terashima et al. (1970) and Hirosawa (1980) that the TNMs are arranged exterior to, and usually between the mesh of, a bed of capillaries.

The distribution of the nerves inside the pit membrane was described by Lynn (1931), Bullock and Fox (1957), de Cock Buning et al. (1981), and Kishida et al. (1982). From the results of dissection, de Cock Buning et al. (1981) estimated that the ophthalmic branch innervated about 12% of the total membrane area, but gave no figures for the other two branches. Only Bullock and Fox (1957) performed degeneration experiments to map the innervation pattern in the pit membrane. Without stating their measuring methods, they reported the areas innervated as "less than one-sixth" of the membrane's area for the ophthalmic branch, "about 60%" for the superficial branch, and "about half the membrane" for the deep branch, "by difference." They did not record the orientation of areas on the membrane.

In contrast, we measured 18% in the dorsocaudal corner for the ophthalmic branch, 36% in the ventral section for the superficial branch, and 46% in the dorsorostral section for the deep branch (Fig. 4). The line of demarcation between each area of innervation was very sharp and clear, and there did not appear to be any overlap between the areas, in agreement with the observations of Bullock and Fox (1957).

We did not measure the absolute size of the individual fibers in the nerve trunks feeding the pit membrane, but the relative size homogeneity in the deep branch, in contrast to the large size variation in the ophthalmic and superficial branches, suggests that the deep branch carries a fairly pure population of infrared fibers and innervates mainly the pit alone, whereas the other two branches innervate, in addition to the pit, other areas of the head and carry, in addition to infrared information from the pit, information of other modalities as well. This is corroborated by the fact that the deep branch innervates the largest portion of the pit membrane, although it contains the smallest number of fibers.

Bullock and Fox (1957) also counted the number of nerve fibers in the trunks supplying the membrane. They counted only aliquot parts of each cross section and corrected to the total cross section in a rattlesnake 965 mm long. They obtained counts of 815, 2,921, and 3,538 for the ophthalmic, superficial, and deep branches, respectively.

On the other hand, we counted all the fibers in each cross section examined, and obtained corresponding counts of 1206, 1832, and 865, respectively. The large difference can be accounted for by the small size of the animals we used, which were about 700 mm in total length, with an average pit membrane diameter of 2.3 mm. According to the figures supplied by Bullock and Fox, a rattlesnake 965 mm long (presumably snout–vent length in their case) would have a pit membrane over 4 mm in diameter and thus have a larger nerve supply because of its greater area. The heavier innervation reported for the rattlesnake could also reflect a greater dependency on use of the pits in capturing warm-blooded prey, whereas our smaller *Agkistrodon* are not so dependent on warm-blooded prey, feeding frequently on amphibians and fish (unpublished observations).

LITERATURE CITED

Bleichmar, H. and E. de Robertis. 1962. Submicroscopic morphology of the infrared receptor of pit vipers. Z. Zellforsch. 56:748–761.

Bullock, T. H. and F. P. J. Diecke. 1956. Properties of an infrared receptor. J. Physiol. (Lond.) 134:47–87.

Bullock, T.H. and W. Fox. 1957. The anatomy of the infrared sense organ in the facial pit of pit vipers. Q. J. Microscop. Sci. 98:219–234.

de Cock Buning, Tj., R. C. Goris and S. Terashima. 1981. The role of thermosensitivity in the feeding behavior of the pit viper *Agkistrodon blomhoffi brevicaudus*. Jpn. J. Herpetol. 9(1):7–27.

De Robertis, E. and H. Bleichmar. 1962. Mitochondriogenesis in nerve fibers of the infrared receptor membrane of pit vipers. Z. Zellforsch. 57:572–582.

Hirosawa, K. 1980. Electron microscopic observations on the pit organ of a crotaline snake *Trimeresurus flavoviridis*. Arch. Histol. Jap. 43:65–77.

Kishida, R., S. Terashima, R. C. Goris and T. Kusunoki. 1982. Infrared sensory neurons in the trigeminal ganglion of crotaline snakes: transganglionic HRP transport. Brain Research 241:3–10.

Kusunoki, T., R. Kishida, T. Kadota and R. C. Goris. 1987. Chemoarchi-tectonics of the brainstem in infrared sensitive and nonsensitive snakes. J. Hirnforsch. 28(1):27–43.

Nachlas, M. M., K.-C. Tsou, E. De Souza,C. S. Chang and A. M. Seligman. 1957. Cytochemical demonstration of succinic dehydrogenase by the use of a new P-nitrophenyl substituted ditertrazol. J. Histochem. Cytochem. 5:420–436.

Ogawa, K. and R. J. Barrnett. 1964. Electron histochemical examination of oxidative enzymes and mitochondria. Nature 203:724–726.

Rosa, C. G. and K.-C. Tsou. 1965. Histochemistry. Demonstration of Sjöstrand membrane particles by an electron cytochemical method. Nature 206:103–105.

Terashima, S., R. C. Goris and Y. Katsuki. 1970. Structure of warm fiber terminals in the pit membrane of vipers. J. Ultrastruct. Res. 31:494–506.

Terashima, S. and R. C. Goris. 1979. Receptive areas of primary infrared afferent neurons in crotaline snakes. Neurosci. 4:1137–1144.

Substance P-like immunoreactivity in the trigeminal sensory nuclei of an infrared-sensitive snake, *Agkistrodon blomhoffi*

Tetsuo Kadota, Reiji Kishida, Richard C. Goris, and Toyokazu Kusunoki

Department of Anatomy, Yokohama City University School of Medicine, Yokohama, Japan

Summary. With the peroxidase-antiperoxidase immunohistochemical method we ascertained the presence of substance P-like immunoreactivity (SPLI) in fibers and cell bodies of the trigeminal sensory system of the pit viper, *Agkistrodon blomhoffi*. There are a few SPLI fibers each in the principal sensory nucleus and the main neuropil of the lateral descending nucleus (i.e., the infrared sensory nucleus); a moderate number in the descending nucleus; and a large number in the caudal subnucleus, the medial edges of the interpolar subnucleus, and the marginal neuropil of the lateral descending nucleus. About 30% of the cell bodies in the ophthalmic and maxillo-mandibular ganglia show SPLI, and of the two craniocervical ganglia, the proximal ganglion has many more cells with SPLI than the distal ganglion. The SPLI distribution in the common trigeminal sensory system is similar to that of mammals, and suggests that the function of this system is also similar. In the infrared sensory system, the differing distribution in the main and marginal neuropils suggests separate functions for these two structures in the system.

Key words: Substance P immunoreactivity – Trigeminal sensory nuclei – V and X ganglia – Infrared sensitive snakes - *Agkistrodon blomhoffi* (Ophidia)

Crotaline snakes have pit organs (infrared sensitive organs) which are richly innervated by heat-receptor fibers of the trigeminal nerve (Bullock and Fox 1957; Terashima et al. 1970; Hirosawa 1980). These snakes have a dual sensory system in the trigeminal complex, the common sensory system and the infrared sensory system (Molenaar 1974; Schroeder and Loop 1976). The common sensory system probably functions as in other animals, but the infrared sensory system has been proven to function only in infrared detection (Terashima and Goris 1977; Kishida et al. 1980; Newman et al. 1980; Gruberg et al. 1984) except at its highest level in the tectum opticum (Hartline et al. 1978; Newman and Hartline 1981). Recently, Kishida et al. (1984) have shown that the lateral descending nucleus of the infrared trigeminal system in the brainstem also receives vagal input to its marginal neuropil, which is thought to affect the processing of infrared information in some way.

Substance P has been shown to be a neuromodulator or neurotransmitter (Pernow 1983), and substance P-like immunoreactivity (SPLI) has been demonstrated in the common trigeminal system of many animal species: mammals (Cuello et al. 1978; Drew et al. 1986; Nomura et al. 1982, 1987), reptiles (Reiner et al. 1984; Wolters et al. 1986), and amphibians (Inagaki et al. 1981; Taban and Cathieni 1983). It has also been demonstrated in the vagal sensory ganglia in mammals and birds (Katz and Karten 1980). In the present study we used immunohistochemical methods to determine the existence and distribution of SPLI in the unique trigeminal sensory system of crotaline snakes, particularly in the primary afferent neurons.

Materials and methods

We used a total of 7 oriental crotaline snakes, *Agkistrodon blomhoffi*. Brain sections were prepared as follows. Under halothane anesthesia we first perfused the animals through the right aortic arch with a phosphate buffer solution (0.1 M, pH 7.4) containing 0.9% NaCl, 5 I.U./ml heparin, and 0.05% of an additional anesthetic, tricaine methanesulfonate, then with Zamboni's fluid (Zamboni and De Martino 1967). The perfusion solutions were kept at 4° C with ice. The brains with the trigeminal ganglia and vagal ganglia attached were removed at once and postfixed overnight at 4° C. The tissues were then placed in the same buffer containing 15% sucrose at 4° C and left in it until they sank (~24 h). Thirty-μm serial frontal sections were cut on a cryostat and mounted alternately on 2 series of gelatin-coated slides. The slides were dried with a fan at room temperature, and then stored in the phosphate buffer containing 0.3% Triton X-100 for at least 4 days.

We performed the immunohistochemical staining according to the peroxidase-antiperoxidase technique of Sternberger et al. (1970) using monoclonal antibody (sera-lab MAS 035, Lot: B3k35, Cuello et al. 1979) against substance P. We obtained satisfactory staining results at a dilution of 1:3000. This reaction was completely blocked by preabsorption of the antibody with 10 μM of substance P (Peptide Institute Inc., Osaka, Japan, 4014-v). After reacting all the slides with the antibody, we counterstained one of the frontal series of each specimen with cresyl violet for cytoarchitectural observation.

To prepare normal material for comparison, we perfused an additional 2 animals with 10% formalin and decalcified the heads in a solution containing 10% formalin and

10% formic acid. Celloidin-embedded heads were cut into a series of 30 μm sections, horizontally in one specimen and frontally in the other. We stained the slides alternately with Klüver-Barrera stain and hematoxylin-eosin.

Results

1. Sensory nuclei

In the trigeminal sensory nuclei of the medulla, only fibers and their varicosities showed substance P-like immunoreactivity (SPLI), staining in various shades of a rusty brown color (hereafter called SPLI fibers). Nowhere in these nuclei do the cell bodies or other structures show this immunoreactivity. All SPLI fibers have varicosities. Some fibers run singly, and others clearly course together in small to medium bundles. It was impossible to observe any particular connections with cell bodies in counterstained Nissl sections.

Common sensory system. As noted by Molenaar (1978a) in other infrared-sensitive snakes, also in our normal material the common sensory trigeminal nucleus can be divided into a principal nucleus and a descending nucleus with 3 subnuclei, i.e., the oralis, interpolaris, and caudalis.

The principal sensory trigeminal nucleus (PrV) in our sections appears delineated by SPLI fibers. Inside the nucleus there are only very few SPLI fibers, some of which clearly enter the nucleus from its periphery. However, no SPLI fibers whatever enter the nucleus from the trigeminal root (Fig. 1).

In the subnucleus oralis, a large number of descending SPLI fibers run along the lateral and dorsal margins of the lateral descending tract of the trigeminal nerve. From these, small bundles of 4–5 fibers each detach themselves to transverse the lateral descending tract and the descending tract of the trigeminal in the direction of the subnucleus oralis. The trajectories of these bundles are all roughly parallel to one another. Some appear to terminate in the subnucleus oralis, and a few pass through the subnucleus oralis to terminate in the underlying reticular formation (Fig. 2).

In the subnucleus interpolaris, descending SPLI fibers continue to run along the dorsolateral margin of the lateral descending tract, with the difference that at this level the number of stained fibers is markedly increased. At various locations along this tract, bundles of fibers detach themselves from the tract, traverse the descending tract and the lateral part of the subnucleus interpolaris, and converge in large numbers on a well-defined region at the medial edge of the subnucleus interpolaris (Fig. 3).

In the subnucleus caudalis, heavily stained SPLI fibers continue to increase in number in the outer margin of the lateral descending tract. A few small bundles branch off to enter the nucleus of the lateral descending tract. Mostly,

however, the nucleus of the lateral descending tract seems to be surrounded on all sides by SPLI fibers which do not enter it. The subnucleus caudalis is situated medially to the nucleus of the lateral descending tract, and 3 different aspects of this subnucleus can be distinguished, according to the level of the sections. In its more rostral aspect, nearly at the level of the obex and therefore still inside the cranium, at its border with the subnucleus interpolaris, a large number of SPLI fibers enter its dorsal portion and continue medially, but there is no lamination, so the structure resembles that of the subnucleus interpolaris. Further caudal, but still inside the cranium, both the lateral and dorsal margins of the subnucleus caudalis are densely occupied by SPLI fibers, and its dorsal and lateral portions are also clearly stained, with lighter staining in the ventromedial portion. A laminated structure is also evident in counterstained sections. Outside the cranium in the spinal cord, the laminated structure continues, but the SPLI staining is lighter, and the nucleus of the lateral descending tract can no longer be seen. The subnucleus caudalis continues caudally to the level of the 3rd cervical vertebra (Figs. 4, 5).

Infrared sensory system. Laterally to the subnucleus caudalis of the common sensory system the lateral descending trigeminal complex (lateral descending tract and its nucleus) forms the infrared sensory system. This complex consists of three layers: the outermost is the marginal neuropil, followed medially by the lateral descending tract proper and then the main neuropil or nucleus of the lateral descending tract. Practically the entire marginal neuropil is stained for SPLI; the lateral descending tract proper shows considerably less staining; and the nucleus of the lateral descending tract contains a number of SPLI fiber bundles, especially in its border regions, with a few of them penetrating the interior of the nucleus. In general, the amount of SPLI fibers in the nucleus of the lateral descending tract is comparable to the very small amount seen in the principal sensory nucleus (Fig. 4).

2. Ganglia

There are 2 ganglia in the trigeminal nerve, the ophthalmic ganglion and maxillo-mandibular ganglion. The vagus nerve is fused with the glossopharygeal hypoglossal and first cervical spinal nerves to form the craniocervical trunk (Auen and Langebartel 1977). In this trunk there are 2 ganglion cell groups, the proximal ganglion and the distal ganglion, but it is impossible to assign any of the component nerves of the trunk specifically to either of these ganglia (Fig. 6).

We found SPLI-positive cell bodies in all of these ganglia. Of the craniocervical ganglia, the proximal has the most SPLI-positive cells; the majority are SPLI-positive

Fig. 1 A–C. Substance P-like immunoreactivity (SPLI) in principal sensory nucleus of trigeminal nerve (*PrV*). **A** Frontal section, × 100. **B** Counterstained neighboring section with trigeminal ganglion (GV) attached. × 13. **C** Diagram of A. *rVs* sensory root of trigeminal nerve; *rVm* motor root of trigeminal nerve; *MV* motor nucleus of trigeminal nerve

Fig. 2 A–C. SPLI in subnucleus oralis of descending trigeminal nucleus (*DVo*). **A** Frontal section. × 100. **B** Counterstained neighboring section. × 13. **C** Diagram of A. *rVIII* root of VIIIth cranial nerve; *dt-dlt* descending tract and lateral descending tract of trigeminal nerve

Fig. 3 A–C. SPLI in subnucleus interpolaris of descending trigeminal nucleus (*DVi*). **A** Frontal section, × 100. **B** Counterstained neighboring section. × 13. **C** Diagram of A

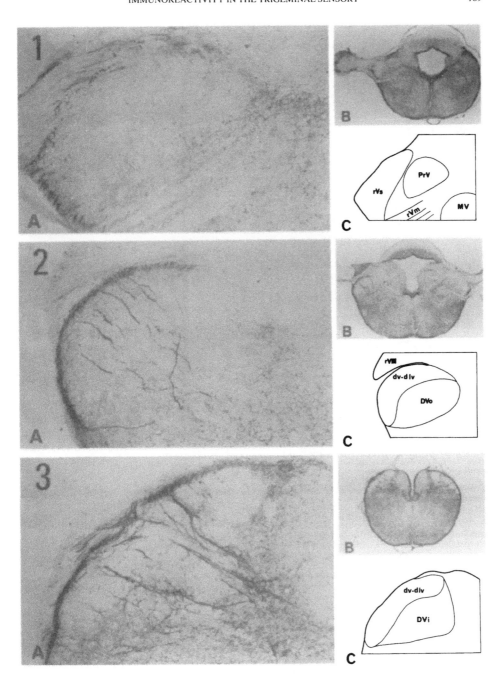

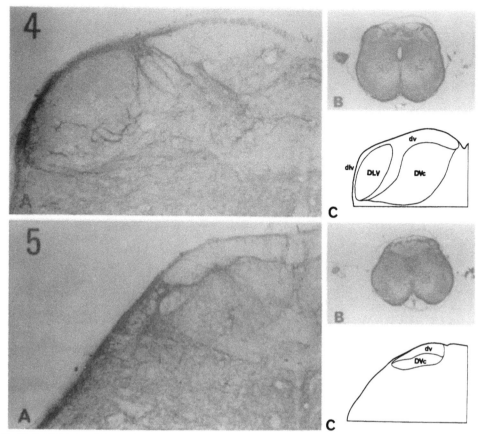

Fig. 4A–C. SPLI in lateral descending nucleus (*DLV*) and rostral part of subnucleus caudalis (*DVc*) of trigeminal nerve. **A** Frontal section, ×100. **B** Counterstained neighboring section. ×13. **C** Diagram of **A**. *dlt* Lateral descending tract of trigeminal nerve; *dt* descending tract of trigeminal nerve

Fig. 5A–C. SPLI in caudal part of subnucleus caudalis of descending trigeminal nucleus (*DVc*). **A** Frontal section, ×100. **B** Counterstained neighboring section. ×13. **C** Diagram of **A**

(Fig. 7C). In contrast, only a few scattered cells in the distal ganglion are positive (Fig. 7D). In the trigeminal nerve. both ganglia show about the same proportion of SPLI-positive cells, i.e., about 30% (Fig. 7A, B). In all 4 ganglia a few SPLI-positive fibers can also be seen.

All SPLI cells show a more or less rounded soma with a single stem process. Both soma and stem are stained with about the same intensity, but it is not possible to distinguish the point at which the stem branches into distal and proximal processes.

Discussion

This report is the first to treat the sensory trigeminal system of crotaline snakes in detail, including the special infrared

sensory system. Meszler et al. (1981) reported that the marginal neuropil of the lateral descending trigeminal complex is made up almost entirely of C fibers. At least a part of the C fibers derive from the vagal nerve (Kishida et al. 1984). The dense SPLI staining in the marginal neuropil noted in the present work can therefore be explained in terms of these C fibers, since it has been shown that a high percentage of C fibers show SPLI (Priestley et al. 1982).

Meszler et al. (1981) also showed that the main neuropil of the lateral descending trigeminal complex is composed mainly of A-delta fibers from the infrared pit organs. In the present work the main neuropil shows very little SPLI activity. which indicates that, at least in the primary crotaline infrared system. A-delta fibers do not have SPLI, and

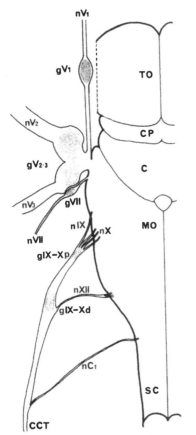

Fig. 6. Sketch showing location of ganglia of cranial nerves V, VII, IX, and X. *C* cerebellum; *CCT* craniocervical trunk; *CP* colliculus posterior; *gV1* ophthalmic ganglion of trigeminal nerve; *gV2–3* maxillo-mandibular ganglion of trigeminal nerve; *gVII* ganglion of facial nerve; *gIX-Xd* distal craniocervical ganglion; *gIX-Xp* proximal craniocervical ganglion; *MO* medulla oblongata; *nC1* first cervical spinal nerve; *nV1* ophthalmic nerve; *nV2* maxillary nerve; *nV3* mandibular nerve; *nVII* facial nerve; *nIX* glossopharyngeal nerve; *nX* vagus nerve; *nXII* hypoglossal nerve; *SC* spinal cord; *TO* tectum opticum

therefore do not employ SP-like substances as neurotransmitters. However, since secondary infrared neurons within the main neuropil send dendrites into the marginal neuropil (Meszler et al. 1981), these dendrites probably receive information via SP-like neurotransmitters.

In a previous report from this laboratory on the distribution of enzyme activity in the brainstem of crotaline snakes, based on succinate dehydrogenase (SDH) and other enzymes as markers, Kusunoki et al. (1987) showed that the main neuropil in the lateral descending trigeminal complex has strong to moderate SDH activity, but its marginal neuropil has only weak SDH activity. The distribution of SPLI fibers in the present study also shows a difference between the main and marginal neuropils. In the light of their nerve connections (trigeminal for the main, vagus for the marginal neuropil), it can therefore be surmised that there is a functional difference between the two, with the main neuropil serving a somatic function and the marginal a visceral function.

There have been many reports on the distribution of SPLI in trigeminal systems of mammals (Hökfelt et al. 1975; Cuello and Kanazawa 1978; Cuello et al. 1978; Del Fiacco and Cuello 1980; Nomura et al. 1982, 1987; Priestley et al. 1982; Salt et al. 1983; Drew et al. 1986), reptiles (Reiner et al. 1984; Wolters et al. 1986), and amphibians (Inagaki et al. 1981; Taban and Cathieni 1983). However, only Drew et al. (1986) carried their description to the level of the subnuclei.

Cuello and Kanazawa (1978), the only workers to describe SPLI in the principal sensory nucleus, reported only isolated groups of SPLI fibers in the principal sensory nucleus. This agrees with the present results. In general, the principal sensory nucleus receives bifurcating ascending fibers that subserve touch and two-point discrimination, and nonbifurcating ascending fibers that subserve touch, two-point discrimination, and position sense (Noback and Demarest 1981). The paucity of SPLI fibers in the principal sensory nucleus indicates that they are probably not involved in these three functions.

Some SPLI fibers are distributed in the subnucleus oralis, and many small bundles transverse the tract and nucleus and terminate in the underlying reticular formation. HRP labeling also shows fiber bundles transversing the tract and nucleus and terminating in the reticular formation in the subnucleus oralis of a lizard species (Barbas-Henry and Lohman 1986). Therefore, these small fiber bundles probably come from the SPLI-positive trigeminal ganglion. Since the subnucleus oralis receives tactile input from the head, mouth, lips, and nose (Noback and Demarest 1981), SPLI fibers are probably involved to a small extent in this function.

There are many SPLI fibers in the medial edge of the subnucleus interpolaris. The dorsomedial part of the subnucleus interpolaris shows strong SDH activity (Kusunoki et al. 1987), and receives fibers from the glossopharyngeal and vagal nerves in pit vipers (Kishida, unpublished data). Therefore, it might play some special role in the sensory trigeminal system with the participation of SPLI fibers. The subnucleus interpolaris receives input from the forehead, cheeks, and angle of the jaw, and from tooth pulp (Noback and Demarest 1981). The medial edge of the subnucleus interpolaris also receives trigeminal projections from the mandibular regions (Molenaar 1978b). Therefore, this area may integrate information from many sensory modalities, possibly from the mouth and pharynx.

The subnucleus caudalis receives the modalities of light touch, pain, and temperature from the entire trigeminal area (Noback and Demarest 1981). The pattern of SPLI fibers within the subnucleus caudalis and the periobex region of the subnucleus interpolaris is similar to the projection pattern of dental nociceptive afferent fibers (Drew et al. 1986). In the mammalian dorsal horn, SP-containing afferents participate in the transmission of mechanical pain (Kuraishi et al. 1985). We have found SPLI fibers in the dental pulp and oral mucosa of pit vipers, and they are

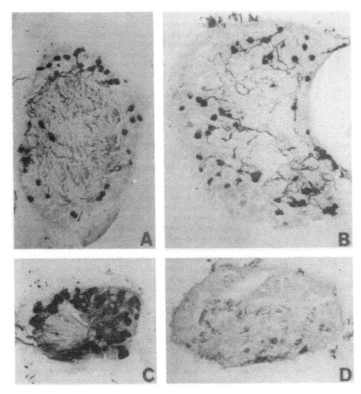

Fig. 7. SPLI in ganglia, frontal
section. A gV1; B gV2–3; C gIX-
Xp; D gIX-Xd. ×100

widely distributed in their skin (unpublished data). Therefore, SPLI-containing trigeminal systems may have the
same functional role as in mammals.

In our material, the glossopharyngeal and vagal nerves
appear fused, which is probably the usûal condition in
snakes (cf. Auen and Langebartel 1977). The fact that SPLI
is abundant in the proximal, but very scarce in the distal
ganglion resembles the condition in birds, where the proxi-
mally located jugular-superior vagal ganglion contains far
more SPLI than the distally located nodose vagal ganglion
(Katz and Karten 1980), but differs from that in mammals,
where both ganglia show large amounts of SPLI (Katz and
Karten 1980). This could reflect the evolutionary status of
reptiles in regard to birds.

References

Auen EL, Langebartel DA (1977) The cranial nerves of the colubrid snake *Elaphe* and *Thamnophis*. J Morphol 154:205–222
Barbas-Henry HA, Lohman AHM (1986) The motor complex and
   primary projections of the trigeminal nerve in the monitor lizard, *Varanus exanthematicus*. J Comp Neurol 254:314–329
Bullock TH, Fox SW (1957) The anatomy of the infrared sense
   organ in the facial pit of pit vipers. Q J Micro Sci 98:219–234
Cuello AC, Kanazawa I (1978) The distribution of substance P

immunoreactive fibers in the rat central nervous system. J
   Comp Neurol 78:129–156
Cuello AC, Del Fiacco M, Paxinos G (1978) The central and pe-
   ripheral ends of the substance P-containing sensory neurones
   in the rat trigeminal system. Brain Res 152:499–509
Cuello AC, Galfre G, Milstein C (1979) Detection of substance
   P in the central nervous system by a monoclonal antibody.
   Proc Natl Acad Sci USA 76:3532–3536
Del Fiacco M, Cuello AC (1980) Substance P- and enkephalin-
   containing neurones in the rat trigeminal system. Neuroscience
   5:803–815
Drew JP, Westrum LE, Ho RH (1986) Mapping of the normal
   distribution of substance P-like immunoreactivity in the spinal
   trigeminal nucleus of the cat. Exp Neurol 93:168–179
Gruberg ER, Newman EA, Hartline PH (1984) 2-Deoxyglucose
   labelling of the infrared sensory system in the rattlesnake, *Cro-
   talus viridis*. J Comp Neurol 229:321–328
Hartline PH, Kass L, Loop MS (1978) Merging of modalities in
   the optic tectum: infrared and visual integration in rattlesnakes.
   Science 199:1225–1229
Hirosawa K (1980) Electron microscopic observations on the pit
   organ of a crotaline snake *Trimeresurus flavoviridis*. Arch Histol
   Jpn 43:65–77
Hökfelt T, Kellerth JO, Nilsson G, Pernow B (1975) Substance
   P: Localization in the central nervous system and in some pri-
   mary sensory neurons. Science 190:889–890
Inagaki S, Senba E, Shiosaka S, Takagi T, Kawai Y, Takatsuki
   K, Sakanaka M, Matsuzaki T, Tohyama M (1981) Regional

distribution of substance P-like immunoreactivity in the frog brain and spinal cord: immunohistochemical analysis. J Comp Neurol 201:243–254

Katz DM, Karten HJ (1980) Substance P in the vagal sensory ganglia: Localization in cell bodies and pericellular arborizations. J Comp Neurol 193:549–564

Kishida R, Amemiya F, Kusunoki T, Terashima S (1980) A new tectal afferent nucleus of the infrared sensory system in the medulla oblongata of crotaline snakes. Brain Res 195:271–279

Kishida R, Yoshimoto M, Kusunoki T, Goris RC, Terashima S (1984) Vagal afferent C fibers projecting to the lateral descending trigeminal complex of crotaline snakes. Exp Brain Res 53:315–319

Kuraishi Y, Hirota N, Sato Y, Hino Y, Satoh M, Takagi H (1985) Evidence that substance P and somatostatin transmit separate information related to pain in the spinal dorsal horn. Brain Res 325:294–298

Kusunoki T, Kishida R, Kadota T, Goris RC (1987) Chemoarchitectonics of the brainstem in infrared sensitive and nonsensitive snakes. J Hirnforsch 28:27–43

Meszler RM, Auker CR, Carpenter DO (1981) Fine structure and organization of the infrared receptor relay, the lateral descending nucleus of the trigeminal nerve in pit vipers. J Comp Neurol 196:571–584

Molenaar GJ (1974) An additional trigeminal system in certain snakes possessing infrared receptors. Brain Res 78:340–344

Molenaar GJ (1978a) The sensory trigeminal systems of a snake in the possession of infrared receptors I. The sensory trigeminal nuclei. J Comp Neurol 179:123–136

Molenaar GJ (1978b) The sensory trigeminal systems of a snake in the possession of infrared receptors II. The central projections of the trigeminal nerve. J Comp Neurol 179:137–152

Newman EA, Hartline PH (1981) Integration of visual and infrared information in bimodal neurons of the rattlesnake optic tectum. Science 213:789–791

Newman EA, Gruberg ER, Hartline PH (1980) The infrared trigemino-tectal pathway in the rattlesnake and in the python. J Comp Neurol 191:465–477

Noback CR, Demarest RJ (1981) The human nervous system Basic principles of neurobiology. McGraw-Hill International Book Company, Tokyo

Nomura H, Shiosaka S, Inagaki S, Ishimoto I, Senba E, Sakanaka M, Takatsuki K, Matsuzaki T, Kubota Y, Saito H, Takase S, Kogure K, Tohyama M (1982) Distribution of substance

P-like immunoreactivity in the lower brainstem of the human fetus; An immunohistochemical study. Brain Res 252:315–325

Nomura H, Shiosaka S, Tohyama M (1987) Distribution of substance P-like immunoreactive structures in the brainstem of the adult human brain: an immunohistochemical study. Brain Res 404:365–370

Pernow B (1983) Substance P. Phamacol Rev 35:85–141

Priestley JV, Somogyi P, Cuello AC (1982) Immunocytochemical localization of substance P in the spinal trigeminal nucleus of the rat: a light and electron microscopic study. J Comp Neurol 211:31–49

Reiner A, Krause JE, Keyser KT, Eldred WD, McKelvy JF (1984) The distribution of substance P in turtle nervous system: a radioimmunoassay and immunohistochemical study. J Comp Neurol 226:50–75

Salt TE, Morris R, Hill RG (1983) Distribution of substance P-responsive and nociceptive neurons in relation to substance P-immunoreactivity within the caudal trigeminal nucleus of the rat. Brain Res 273:217–228

Schroeder DM, Loop MS (1976) Trigeminal projections in snakes possessing infrared sensitivity. J Comp Neurol 169:1–14

Sternberger LA, Hardy PH Jr, Cuculis JJ, Meyer HG (1970) The unlabeled antibody enzyme method of immunohistochemistry: Preparation and properties of soluble antigen-antibody complex (horseradish peroxidase-antihorseradish peroxidase) and its use in identification of spirochetes. J Histochem Cytochem 18:315–333

Taban CH, Cathieni M (1983) Distribution of substance P-like immunoreactivity in the brain of the newt (*Triturus cristatus*). J Comp Neurol 216:453–470

Terashima S, Goris RC (1977) Infrared bulbar units in crotaline snakes. Proc Jpn Acad Ser B 53:292–296

Terashima S, Goris RC, Katsuki Y (1970) Structure of warm fiber terminals in the pit membrane of vipers. J Ultrastruct Res 31:494–506

Wolters JG, Ten Donkelaar HJ, Verhofstad AAJ (1986) Distribution of some peptides (substance P, [Leu]enkephalin, [Met]enkephalin) in the brain stem and spinal cord of a lizard, *Varanus exanthematicus*. Neuroscience 18:917–946

Zamboni L, De Martino C (1967) Buffered picric acid-formaldehyde: A new, rapid fixative for electron microscopy. J Cell Biol 35:148A

SUBSTANCE P-LIKE IMMUNOREACTIVE FIBERS IN THE TRIGEMINAL SENSORY NUCLEI OF THE PIT VIPER, *TRIMERESURUS FLAVOVIRIDIS*

S. Terashima

Department of Physiology, University of the Ryukyus School of Medicine, Nishihara-cho,
Okinawa 903-01, Japan

Abstract—Substance P-like immunoreactive nerve fibres were located in the trigeminal sensory system of the infrared-sensitive snake, *Trimeresurus flavoviridis*, using the immunohistochemical method. There are two trigeminal sensory systems in the medulla of this animal: the descending nucleus and the lateral descending nucleus. The descending nucleus is equivalent to the trigeminal spinal nucleus in other vertebrates, and the lateral descending nucleus is a special trigeminal sensory nucleus belonging to the infrared sensory system.

In the present study we determined that the lateral descending nucleus is completely ensheathed by large numbers of substance P-like immunoreactive fibers. The distribution of these fibers seems to be similar to that of the thin vagal unmyelinated fibers, rather than to that of the thick trigeminal myelinated fibers. More substance P-like immunoreactive nerve fibers were observed in the lateral descending tract than in the descending tract. Almost no dense substance P-like immunoreactive fibers were found in these tracts rostral to the lateral descending nucleus or rostral to the subnucleus caudalis of the descending nucleus. The substance P-like immunoreactive fibers in the lateral descending tract extended to those of Lissauer's tract of the spinal cord, and the substance P-like immunoreactive fibers surrounding the Lissauer's tract were similar in appearance to those of the lateral descending nucleus. This nucleus seems to have developed from the elements existing in Lissauer's tract, and also to have a similar modulating function. The primary nucleus of the infrared sensory system is the most substance P-like immunoreactive nucleus in the trigeminal sensory system of this animal. Even in the trigeminal sensory system, substance P-like immunoreactive fibers seem not to be related solely to the nociceptive sensation.

The nucleus descendens of the trigeminal nerve (DV), which functions mainly for skin sensations of the facial area, is located in the medulla oblongata of vertebrates. This nucleus is divided into three parts: the subnucleus oralis (DVo), the interpolaris (DVi), and the caudalis (DVc).[3,20] Some of the DVc function is believed to be related to nociception. If the DVc is related to the information processing of the nociceptive sensation, it is conceivable that there could be a set of specific neurotransmitters for this purpose. One of the putative neurotransmitters could be substance P (SP), although a variety of SP functions on various nerve tissues has been reported (e.g. Refs 19 and 21).

The trigeminal sensory system in the medulla oblongata of the crotaline snake has two parts: the descending nucleus (DV) and the lateral descending nucleus (DLV).[17,23] The structure of the DV is similar to that of other vertebrates, and can be divided into three subnuclei.[18] This nucleus is considered equivalent to that of other animals which functions for the same skin sensation. The DLV, however, which is located just lateral to the DV, is specific to infra-red sensing animals and is solely related to the infra-red sensory processing.[24,26] The results of an histological experiment on transganglionic horseradish peroxidase (HRP) transport[11] are also consistent with the physiological observation.

In the present study we investigated dense substance P-like immunoreactive (SP-LI) fiber distribution in various parts of the infra-red sensory system of the Asian pit viper, *Trimeresurus flavoviridis*, and then compared the distribution in the DV and the DLV each of which functions in separate modalities. A preliminary report was previously published elsewhere.[8]

Abbreviations: DAB, 3,3'-diaminobenzidine; dlv, tractus descendens lateralis of the trigeminal nerve; DLV, nucleus descendens lateralis of the trigeminal nerve; dv, tractus descendens of the trigeminal nerve; DV, nucleus descendens of the trigeminal nerve; DVc, subnucleus caudalis of the nucleus descendens of the trigeminal nerve; HRP, horseradish peroxidase; PAP, horseradish peroxidase–antihorseradish peroxidase; PB, phosphate buffer; PBS, phosphate-buffered saline; PBST, phosphate-buffered saline containing 0.3% Triton X-100; RC, nucleus reticularis caloris; SP, substance P; SP-LI, substance P-like immunoreactivity.

EXPERIMENTAL PROCEDURES

For the purpose of detecting SP-LI fibers in the brainstem of pit viper, *Trimeresurus flavoviridis*, we used six animals, including both young and adult animals, ranging in weight from 15 to 100 g or more. Three of them were for transverse sectioning, two were for horizontal sectioning, and one was for sagittal sectioning. The sections were treated using the horseradish peroxidase–antihorseradish peroxidase (PAP) immunohistochemical method of Sternberger *et al.*[25]

The animals were anesthetized with halothane, and were then perfused through branches of the aorta which lead to the head with cold 0.75% saline followed by 100 ml of 4°C

Zamboni's fixative.[28] The brains, ganglia, and pit membranes were dissected free, and postfixed in the same fixative at the same temperature for 48 h. They were left in 10% sucrose for two days, and then placed in 20% sucrose for 2 days. They were then immersed for 24 h at 37–40°C in phosphate buffer (PB) containing gelatin, and then cooled by cold water. The gelatin-embedded blocks were quickly frozen in isopentane chilled by dry ice. They were cut on a cryostat into 20-μm free-floating sections.

These free-floating sections were incubated in phosphate-buffered saline (PBS)[6] for 1 h, then treated in H_2O_2 to block the peroxidase in the red cells during subsequent treatment with 3,3'-diaminobenzidine (DAB). They were then washed in PBS containing 0.3% Triton X-100 (PBST) for 15 min, and placed for reaction in normal goat serum diluted (1:40) with PBST, and rinsed in PBST for 30 min. The sections were incubated in anti-SP rabbit serum (1:1600; Immuno Nuclear Corporation) for 24–36 h at 4°C. They were then rinsed in PBST for 30 min, and incubated for 1 h at room temperature in goat antirabbit IgG serum (1:200; Cappel). Thereafter, the sections were washed in PBST for 30 min and incubated for 1 h at room temperature in PAP complex (1:200, Cappel). They were then rinsed in PBST for 30 min, soaked in 0.02% DAB containing 0.006% H_2O_2 for 13 min and in some cases counterstained for 10 min at room temperature with 0.005% OsO_4.

To assure the specificity of our immunoreaction to SP-LI, tests were carried out as follows: staining was completely eliminated (1) by pretreatment of SP-antibody with 100 μg or more of SP, (2) by elimination of SP-antibody, and (3) by elimination of IgG.

We applied colchicine in the fouth ventricle of some specimens in order to enhance SP-LI soma detection.

In order to locate the origin of SP-LI nerve fibers, ganglionectomy was performed on 10 specimens. In five of the specimens, it was for the trigeminal nerve unilaterally and in the other five for the vagus and glossopharyngeus. These animals were then killed for immunohistological investigation for SP-LI in five days to three weeks after the operation.

RESULTS

Dense distribution of SP-LI fibers in the medulla oblongata was limited primarily to the DV + dv, the DLV + dlv, and the vagal complex. Less dense distributions of SP-LI fibers were found in the periaqueductal area, the periventricular area of IV, and the reticular formation. However, in the present investigation we examined the SP-LI fiber distribution in the infra-red sensory system which includes the pit membrane, the trigeminal ganglion, the DLV, the nucleus reticularis caloris (RC) and the tectum. Following that procedure, we compared the DLV with the DV, and the trigeminal ganglion with the nodose ganglion.

We found SP-LI somata in the ganglia of the trigeminal and vagal nerve, but could not find them in the medulla oblongata even with colchicine treatment.

Pit membrane, trigeminal nerve trunk and trigeminal ganglion

No SP-LI fibers were found in the pit membrane. They were found in the trigeminal nerve trunk and ganglion (Fig. 1A). Many SP-LI somata were observed in the ganglion.

Nucleus descendens lateralis plus tractus descendens lateralis of the trigeminal nerve

SP-LI fibers were found to be surrounding the DLV in all sections of the three different dimensions: transverse sections (Fig. 2A and B), sagittal sections (Figs 2D and 3A), and a horizontal section (Fig. 2E). The SP-LI fibers were invading the DLV, where their branches were twisting and spreading in no particular direction. In the tractus descendens lateralis of the trigeminal nerve (dlv), especially dorsolateral surface of the medulla, dense SP-LI fibers were observed (Fig. 2A and B). SP-LI fibers in the dlv extended to the SP-LI fibers of Lissauer's tract of the spinal cord (Fig. 2C).

Nucleus reticularis caloris

There were some SP-LI fibers in the reticularis caloris (RC) which were the same as those in the surrounding reticular formation, so there was no difference in the density of the RC and its surroundings.

Tectum opticum

Dense SP-LI fibers were absent in layer 6 (stratum album centrale). Furthermore, SP-LI fibers were neither in the lamina 7b–13 where visual responses were recorded, nor in the laminae 7a and 7b where infra-red cells were found.[9]

Nucleus descendens and tractus descendens of the trigeminal nerve

SP-LI fibers were located abundantly in the tractus descendens of the trigeminal nerve (dv), and some fibers were invading the nucleus, while the others were transversing the DVc to the solitary tract (Fig. 2F). The difference in SP-LI fiber density among the three subnuclei of the DV was noticeable: only the DVc was dense in SP-LI fiber distribution (Fig. 3B and C).

Vagal ganglion and vagal nerve trunk

There were SP-LI fibers in the vagal nerve root (Fig. 1B) and trunk (Fig. 1D). The number of SP-LI negative-fibers seemed to be greater than the number of SP-LI fibers. SP-LI cells were also found in the ganglion (Fig. 1C).

Solitary tract

The SP-LI fibers passed through the DVc to the solitary tract (Fig. 2F), where a dense SP-LI fiber bundle was observed.

Spinal cord

SP-LI fibers were densely located on the surface of Lissauer's tract of the cervical spinal cord (Fig. 2C) and were surrounding the tract. The fibers seemed to be distributed in a pattern equivalent to the DLV (Fig. 2A and B). The SP-LI fibers in this tract extended to the SP-LI fibers in the dv and dlv.

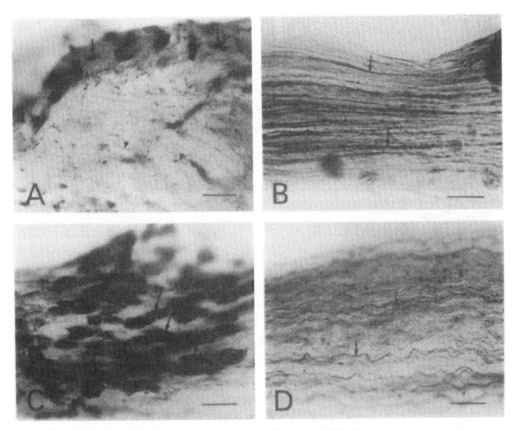

Fig. 1. Photomicrographs of wholemount preparations illustrating SP-LI in the trigeminal and the vagal nerve. (A) SP-LI somata and nerve fibers of the trigeminal ganglion (a part of the maxillary division); (B) the vagal nerve root; (C) SP-LI cells in the nodose ganglion; (D) the vagal nerve trunk. Arrows in (A) and (C) indicate SP-LI somata, and in (B) and (D) indicate SP-LI fibers. Arrowheads in (A) indicate SP-LI fibers. Left side is central for (B)–(D). Bars = 50 μm in (A), 25 μm in (B), 30 μm in (C), and 100 μm in (D).

Effects of ganglionectomy

After the trigeminal nerve ganglionectomy, drastic reduction of SP-LI content was observed in the medulla oblongata. However, SP-LI in the dv and dlv at the level of the DVc and the DLV remained unchanged.

After the glossopharyngeus and the vagus ganglionectomy, no change was observed in the distribution of SP-LI in the medulla oblongata.

DISCUSSION

Many of the SP-LI fibers crossed the DVc and were bound for the solitary tract (Fig. 2F), so they seemed to be visceral afferent. Similar fibers were also reported in rats.[15] There were SP-LI neurons in the vagal sensory ganglia of birds and mammals which were perhaps acting as visceral afferent from baroreceptors or other receptors.[7,10]

There are myelinated and unmyelinated fibers in the dlv of infra-red-sensitive snakes,[16] and the tract is

separated into two layers; the inner for myelinated thick fibers, and the outer for unmyelinated thin fibers of which terminals form marginal neuropil.[12] Present results indicate that SP-LI fibers are localized mainly in the part containing unmyelinated thin fibers.

The SP-LI cells in the trigeminal ganglion and a number of SP-LI fibers are reported in other animals,[13,27] and it has also proved to be true in this animal (Fig. 1A). But it is interesting that SP-LI fibers of the DV + dv and the DLV + dlv seemed more likely to be related to the vagus than to the trigeminal nerve. This is suggested by the following: (1) the general distribution of SP-LI fibers around the DLV (Fig. 2A) was similar to that of unmyelinated fibers from the vagus which were revealed by the transganglionic HRP technique;[12] (2) the area of greatest SP-LI fiber density was found posterior to the DLV (Fig. 2E) at which level the vagal nerve has its roots, while only a few dense bands of SP-LI fibers were observed rostral to the DVc (Fig. 3B and C); (3)

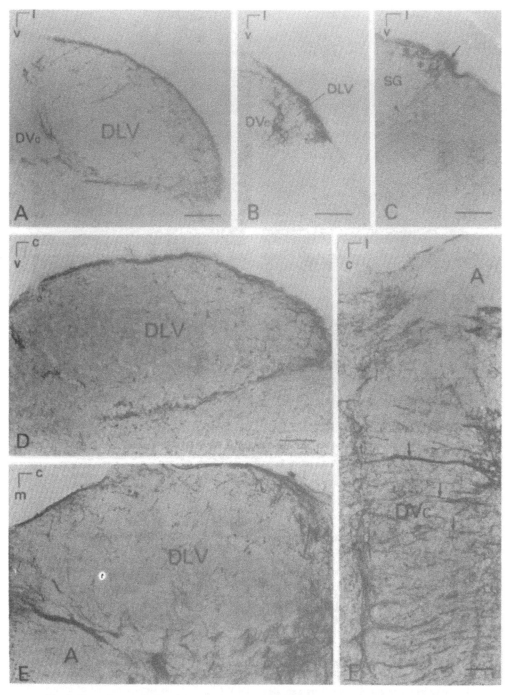

Fig. 2. Photomicrographs illustrating SP-LI fibers in the DLV and the DVc of pit viper, *Trimeresurus flavoviridis*. (A) transverse section of the DLV; (B) transverse section of the caudal end of the DLV; (C) transverse section of Lissauer's tract of the cervical spinal cord; (D) sagittal section of the DLV; (E) horizontal section of the DLV; (F) horizontal section of the DVc. In (C) asterisks indicate areas of the tract surrounded by the SP-LI fibers (arrow), and an arrowhead indicates SP-LI fibers running to the commissura cornuae dorsalis. In (F) arrows indicate SP-LI fibers transversing the DVc to solitary tract (st). Dense SP-LI fibers are surrounding the DLV (A, B, D and E). Abbreviations: A, acoustic area; SG, substantia geratinosa; c, caudal; l, lateral; m, medial; v, ventral. Bars = 100 µm in all illustrations.

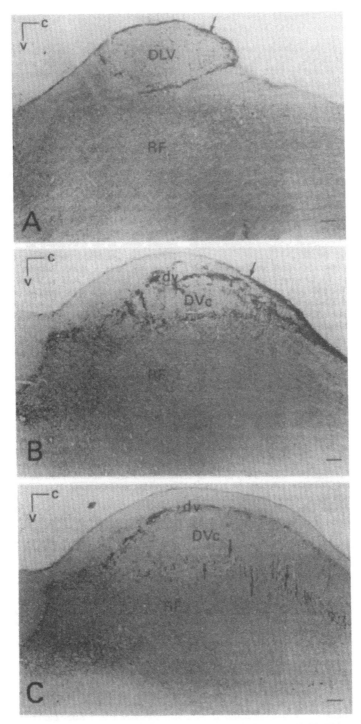

Fig. 3. Photomicrographs illustrating SP-LI fibers in the DLV and DVc. (A) Sagittal section through the DLV; (B) sagittal section of the DVc through the border between the DLV and DVc; (C) sagittal section of the DVc through the center; enlarged DLV in (A) is sited in Fig. 2D. Arrows indicate SP-LI fibers in the dlv (A) and (B), and in the DVc (C). Abbreviations: RF, reticular formation; c, caudal; v, ventral. Bars = 100 µm in all illustrations.

ganglionectomy of the trigeminal nerve drastically reduced SP-LI in the dv, while it did not affect the density of SP-LI around the DLV.

The results of the ganglionectomy of the vagal and glossopharyngeal nerve in the present research do not prove SP-LI depletion in the medulla oblongata. Judging from other data[7] we could not expect complete SP-LI depletion in this area by ganglionectomy. We could not find SP-LI somata in the medulla oblongata of our material even with colchicine treatment, although they are reported to be found in other animals.[14,22] Some SP-LI compensation mechanism would have come from the remaining nerve terminals and cells are found in SP-LI in goldfish.[5] It seems to us that in order to detect a slight depletion of SP-LI after vagal ganglionectomy in our material we might need to utilize a more sensitive method such as radioimmunoassay for SP.[7]

It has been reported that spinal ganglion cells send out SP-LI fibers to Lissauer's tract in rats and guinea-pigs.[1] It has also been reported that in monkeys, all parts of Lissauer's tract contain very large numbers of primary afferent axons.[2] The origin of the SP-LI fibers of Lissauer's tract in our material might also be postulated as being in the spinal ganglion cells.

It is already known that in other animals, SP-LI, which is located lateral to the dv, extends caudally with a band of SP-LI fibers.[22] The results of our present investigation show that SP-LI fibers in the dv and the dlv were together ensheathing the DLV, and were continuous to Lissauer's tract in the spinal cord. The characteristic distribution of the SP-LI fibers ensheathing the DLV was still maintained in Lissauer's tract.

Just inside the unmyelinated fiber part within the dlv there is a zone (marginal neuropil) for synaptic connection between unmyelinated fiber terminals and dendrites of intrinsic cells.[16] These synaptic connections are different from the main neuropil of the trigeminal infra-red primary afferent which makes synaptic contact with dendrites of large secondary relay cells in the nucleus. So, it seems that the SP-LI fiber bundle in the dlv is not a mere part of a tract, but rather forms the site for a function such as modulating for some sense, most probably for the warmth sense.

It is also probable that Lissauer's tract has marginal neuropil, although this has not been proved in our present data. The DLV possibly developed from Lissauer's tract which might include the origin of the secondary relay neurons, and might share the comparable warmth sensing function with the similar modulating mechanism of sensation.

SP is thought to be a transmitter related to the nociceptive sensation, although it does not seem to work exclusively for that sensation (e.g. Ref. 4). Our results suggest the possibility that even in the trigeminal sensory system, SP-LI fibers would not be related solely to the nociceptive sensation. It is also conceivable that SP-LI fibers from the vagal nerve might not have a direct function in nociceptive sensation, but rather work as a modulator as proposed to be the case in the infra-red sensory system. It may be postulated that some kind of visceral afferent input interacts with somatosensory input in the DVc. This also seems true of the dorsal horn in the spinal cord, and such an interaction would be a factor in causing referred pain.

Although there are some similarities in nuclear organization, such as both the DV and the DLV having similar somatotopical organization,[11,18] some histological differences still exist. For example, the DLV is surrounded entirely by the SP-LI fibers, while the laminated DVc[18] is not. Functional differences in information processing seem to be reflected in these results. No strong relationship to dense SP-LI such as that found in the DLV was found in the rest of the infra-red sensory system from the receptor level to the central level (tectum).

Acknowledgements—We gratefully acknowledge the valuable advice received concerning our microphotography from Dr S. Tanaka, Professor of Anatomy of the University of the Ryukyus. We also thank Mr H. Uehara for his technical assistance.

REFERENCES

1. Barber R. P., Vaughn J. E., Slemmon J. R., Salvaterra P. M., Roberts E. and Leeman S. E. (1979) The origin, distribution and synaptic relationships of substance P axons in rat spinal cord. *J. comp. Neurol.* **184**, 331–352.
2. Coggeshall R. E., Chung K., Chung J. M. and Langford L. A. (1981) Primary afferent axons in the tract of Lissauer in the monkey. *J. comp. Neurol.* **196**, 431–442.
3. Crosby E. C. and Yoss R. E. (1954) The phylogenetic continuity of neural mechanisms as illustrated by the spinal tract of V and its nucleus. *Res. Publs Res. Ass. nerv. ment. Dis.* **33**, 147–208.
4. Dubner R. and Bennett G. J. (1983) Spinal and trigeminal mechanisms of nociception. *Ann. Rev. Neurosci.* **6**, 381–418.
5. Finger T. E. (1984) Vagotomy induced changes in acetylcholinesterase staining and substance P-like immunoreactivity in the gustatory lobes of goldfish. *Anat. Embryol.* **170**, 257–264.
6. Hartman B. K. (1973) Immunofluorescence of dopamine-hydroxylase: application of improved methodology to the localization of the peripheral and central noradrenergic nervous system. *J. Histochem. Cytochem.* **21**, 312–332.
7. Helke C. J., O'Donohue T. L. and Jacobowitz D. M. (1980) Substance P as a baro- and chemoreceptor afferent neurotransmitter: immunocytochemical and neurochemical evidence in the rat. *Peptides* **1**, 1–9.
8. Kadota T., Kishida R., Goris R. C., Kusunoki T. and Terashima S. (1984) Substance P-immunoreactive fibers in the trigeminal sensory nuclei of snakes with infrared sensitivity. *Neurosci. Res. Neradn. Suppl.* **1**, S99.
9. Kass L., Loop M. S. and Hartline P. H. (1978) Anatomical and physiological localization of visual and infrared cell layers in tectum of pit vipers. *J. comp. Neurol.* **182**, 811–820.

10. Katz D. M. and Karten H. J. (1980) Substance P in the vagal sensory ganglia: localization in cell bodies and pericellular arborizations. *J. comp. Neurol.* **193**, 549–564.
11. Kishida R., Terashima S., Goris R. C. and Kusunoki T. (1982) Infrared sensory neurons in the trigeminal ganglia of crotaline snakes: transganglionic HRP transport. *Brain Res.* **241**, 3–10.
12. Kishida R., Yoshimoto M., Kusunoki T., Goris R. C. and Terashima S. (1984) Vagal afferent C fibers projecting to the lateral descending trigeminal complex of crotaline snakes. *Expl Brain Res.* **53**, 315–319.
13. Lehtosalo J. I., Uusitalo H., Stjernschantz J. and Palkama A. (1984) Substance P-like immunoreactivity in the trigeminal ganglion. *Histochemistry* **80**, 421–427.
14. Ljungdahl Å., Hökfelt T. and Nilsson G. (1978) Distribution of substance P-like immunoreactivity in the central nervous system of the rat. I. Cell bodies and nerve terminals. *Neuroscience* **3**, 861–943.
15. Lorez H. P., Haeusler G. and Aeppli L. (1983) Substance P neurones in medullary baroreflex areas and baroreflex function of capsaicin-treated rats: comparison with other primary afferent systems. *Neuroscience* **8**, 507–523.
16. Meszler R. M., Auker C. R. and Carpenter D. O. (1981) Fine structure and organization of the infrared receptor relay, the lateral descending nucleus of the trigeminal nerve, in pit vipers. *J. comp. Neurol.* **196**, 571–584.
17. Molenaar G. J. (1974) An additional trigeminal system in certain snakes possessing infrared receptors. *Brain Res.* **78**, 340–344.
18. Molenaar G. J. (1978) The sensory trigeminal system of a snake in the possession of infrared receptors. *J. comp. Neurol.* **179**, 123–136.
19. Nicoll R. A., Schenker C. and Leeman S. E. (1980) Substance P as a transmitter candidate. *A. Rev. Neurosci.* **3**, 227–268.
20. Olszewski J. (1950) On the anatomical and functional organization of the spinal trigeminal nucleus. *J. comp. Neurol.* **92**, 401–413.
21. Otsuka M. and Konishi S. (1983) Substance P—the first peptide neurotransmitter? *Trends Neurosci.* August, 317–320.
22. Reiner A., Krause J. E., Keyser K. T., Eldred W. D. and McKelvy J. F. (1984) The distribution of substance P in turtle nervous system: a radioimmunoassay and immunohistochemical study. *J. comp. Neurol.* **226**, 50–75.
23. Schroeder D. M. and Loop M. (1976) Trigeminal projections in snakes possessing infrared sensitivity. *J. comp. Neurol.* **195**, 477–500.
24. Stanford L. R. and Hartline P. H. (1980) Spatial sharpening by second-order trigeminal neurons in crotaline infrared system. *Brain Res.* **18**, 115–123.
25. Sternberger L. A., Hardy P. H., Cuculis J. J. and Meyer H. G. (1970) The unlabeled antibody enzyme method of immunohistochemistry: preparation and properties of soluble antigen–antibody complex (horseradish peroxidase–antihorseradish peroxidase) and its use in identification of spirochetes. *J. Histochem. Cytochem.* **18**, 315–333.
26. Terashima S. and Goris R. C. (1977) Infrared bulbar units in crotaline snakes. *Proc. Japan Acad. Ser. B.* **53**, 292–296.
27. Tervo K., Tervo T., Eränkö L., Eränkö O. and Cuello A. C. (1981) Immunoreactivity for substance P in the Gasserian ganglion, ophthalmic nerve and anterior segment of the rabbit eye. *Histochem. J.* **43**, 435–443.
28. Zamboni L. and de Martino C. (1967) Buffered picric acid-formaldehyde: a new rapid fixative for electron microscopy. *J. Cell Biol.* **35**, 148A.

Department of Anatomy, Yokohama City University School of Medicine, Yokohama, Japan

Chemoarchitectonics of the Brainstem in Infrared Sensitive and Nonsensitive Snakes

By Toyokazu Kusunoki, Reiji Kishida, Tetsuo Kadota, and Richard C. Goris

With 9 Figures and 2 Tables

Summary: The crotaline snake *Agkistrodon* possesses infrared receptors, whereas the colubrid *Elaphe quadrivirgata* does not. We compared the histochemical activity of succinate dehydrogenase (SDH), monoamine oxidase (MAO), and acetylcholinesterase (AChE) in the brainstem of these 2 species, by the method of Nachlas et al. (1957), Glenner et al. (1957), and Koelle and Friedenwald (1949), respectively, and made the following observations.

Visual system: The tectum opticum (TO) exhibited strong or moderate AChE and SDH activity in areas receiving retinal projections, i.e. the str. zonale (sz), str. fibrosum et griseum superficiale (sfgs), and narrow areas between small tight fasciculi of the tr. opticus. The sfgs was divided into 2 sublayers, a superficial and a deep, by the intensity of AChE activity. The deep sublayer of the sfgs and sfc of *Agkistrodon* were stained more strongly than other layers. Numerous fibers within the TO showed MAO activity. The entire sfgs of *Agkistrodon* was thinner than in *Elaphe*. The nucl. posterodorsalis showed moderate AChE, and weak SDH and MAO activity in *Agkistrodon*, but lack of AChE, weak SDH, and moderate MAO activity in *Elaphe*.

Infrared system: This system was present only in *Agkistrodon*. The nucl. of the lateral descending trigeminal tract (dIV) and the nucl. reticularis caloris (rc) showed strong to moderate SDH activity in the main neuropil and/or perikarya. These nuclei were not conspicuous in AChE preparations. The marginal neuropil of the dIV had weak SDH, and moderate AChE and MAO activity.

Common sensory trigeminal system: Moderate activity of the 3 enzymes was seen in the nucl. tr. descendens n. trigemini (dl). In the dorsomedial part of the nucl. interpolaris, the round limited portion was stained strongly for SDH and AChE. Cells of the nucl. tr. mesencephalicus n. trigemini showed strong SDH and AChE activity.

Other regions: In *Elaphe*, there was strong to moderate AChE and SDH activity in the nucl. of the fasciculus longitudinalis medialis, nucl. centralis superior, raphe nuclei, and reticular nuclei, but only weak activity in *Agkistrodon*. We also found the following similarities in the 2 species. Strong to moderate AChE and SDH activity was observed in the motor nuclei of the cranial nerves, pretectal nuclei excepting the nucl. posterodorsalis, nucl. opticus basalis, and nucl. posterolateralis tegmentalis. Strong to moderate activity of the 3 enzymes together was detected in the nucl. interpeduncularis as found in other animals previously studied, and in the nucl. commissurae cornae dorsalis, nucl. cochlearis angularis, and the molecular and granular layer of the cerebellum. The core of the tr. retroflexus was stained strongly for SDH. A tract and a nucleus, which seemed to correspond to the tr. interpedunculotegmentalis and the nucl. tegmentalis dosalis of mammals, were characterized by plentiful AChE activity. Thus the two snakes studied showed the same general SDH and/or AChE activity in the somatic nuclei as is seen in other vertebrates. MAO activity in these two snakes was found diffusely in the somatic neuropil as well as the visceral neuropil, and in most fibers in the brainstem. The differences observed might be ascribed to the differing ecology of the 2 species.

Introduction

Snakes with pit organs (some Boidae and all Crotalinae) use infrared information to detect prey (De Cock Buning et al., 1981; De Cock Buning, 1983). Special nuclei for the infrared sensory system have been demonstrated in their brainstem by HRP and Fink-Heimer's method. These are the nucl. descendens lateralis n. trigemini (Molenaar, 1974, 1978a, b; Schroeder and Loop, 1976; Kishida et al., 1980) and the nucl. reticularis caloris, which is not found in the Boidae (Gruberg et al., 1979; Kishida et al., 1980; Newman et al., 1980). It is of interest to study the chemoarchitectonical differences in the brainstem of snakes which possess and snakes which do not possess these organs, because physical characters may be reflected in brain structures as the representation of histochemical enzymatic activity. For this purpose we studied *Agkistrodon*, which has a pair of pit organs, and *Elaphe*, which lacks them entirely.

Materials and methods

Ten brains each of active (i.e., not hibernating) *Agkistrodon blomhoffi* (Crotalinae) and *Elaphe quadrivirgata* (Colubridae) were cut into fresh frozen sections of 30 μm thickness. These were stained with Nachlas' method for succinate dehydrogenase (SDH) (Nachlas et al., 1957), Glenner's method for monoamine oxidase (MAO) (Glenner et al., 1957), and Koelle and Friedenwald's method for acetylcholinesterase (AChE) (Koelle and Friendenwald, 1949). Each group of sections was incubated at 37°C at pH 7.6 for 20–30 minutes; 37°C at pH 7.6 for 40 minutes; and 37°C at pH 6.6 and DFP⁻⁷ for 2 hours, respectively. For histological identifi-

183

cation, frontal, sagittal, and horizontal serial sections of the brains of both species were stained with KLÜVER-BARRERA's method or Bodian staining as modified by OTSUKA (1962).

Results

SDH and AChE activity was seen in the perikarya and neuropil, and MAO activity was found mainly in the neuropil and in many fibers within the brainstem. AChE activity was also seen in the walls of capillaries, but will not be considered in this report. The regional distribution of the 3 enzymes is shown in Table 1 (*Agkistrodon*) and Table 2 (*Elaphe*). The main differences and similarities can be stated as follows.

Pretectum

The nucl. lentiformis mesencephali showed SDH and MAO activity in the neuropil, and there were a few SDH-positive perikarya in both species (Fig. 2-A).

Prominent differences in AChE distribution were found in the nucl. posterodorsalis: AChE activity in the neuropil was strong in *Agkistrodon*, but negligible in *Elaphe* (Fig. 1-A).

Tectum opticum (TO)

Macroscopically, it seemed that the TO of *Elaphe* was more developed than that of *Agkistrodon*. Although the chemoarchitectonical stratifications in the TO were indistinct, several layers could be observed, as shown in Tables 1 and 2. Generally, enzymatic activity was stronger in *Elaphe* than in *Agkistrodon*, and there were some differences in the stratum fibrosum et griseum superficiale (sfgs) and the stratum griseum centrale (sgc). Two sublayers, a superficial and a deep, could be distinguished by the intensity of AChE activity in the sfgs of both species (Fig. 9). The deep sublayer stained more strongly for AChE than the

Table 1. AChE, SDH, and MAO activity in *Elaphe*.

f, fibers; n, neuropil; p, perikarya; per, periphery; L, large cells; M, medial part; R, rostral part; RM, rostromedial part; S, a small number; 0, negligible; 1, weak; 2, moderate; 3, strong; *, note.

regions	AChE	SDH	MAO
MESENCEPHALON			
nucl. lentiformis mesencephali	n1, p1	n2, p2(L)	n2
nucl. geniculatus pretectalis	n1, p2	n2	n2
nucl. posterodorsalis	0	n1	n2
nucl. pretectalis	n1,	n3	n2, f2
paratorus	p1	n3, p2	n1, f1
nucl. lat. profundus mesencephali	n0−1, p0−1	p1(S)	f1
nucl. profundus mesencephali	p3	n1, p2	f1
nucl. opticus basalis	n1−2, p2	n2, p3	f1
nucl. posterolateralis tegmentalis	n3, p3	n1, p1	n1
nucl. ruber	n1−2, p1	n1, p1	f1
nucl. isthmi	n2, p1	n1, p1	n2
nucl. of fasc. longitudinalis med.	p3	0	n3(?)
nucl. interpeduncularis			
anterolateral part	n0, p1	n3	n3
medial part	n3(R), p1	n3	n3
neuropil of nucl. interpeduncularis	n3(RM)	n3	n3
nucl. tr. mesencephalicus n. V	p3	p3	0
central gray	n0−1, p1	n2	n2, p1(?)
optic tectum			
str. zonale	n2, p1	n2	n1
str. opticum	0*	0*	f1
str. fib. gris. sup. (superf.)	n1−2	n3, p1(S)	n2
str. fib. gris. sup. (deep)	n2−3	n3, p1(S)	n2
str. griseum centrale	n1−2**	n1, p1(S)	n1
str. fibrosum centrale	n1, p1	n1, p1	f2
str. griseum periventriculare	n1, p1	n1−2, p1	n2
str. ependymale	0	0	1
RHOMBENCEPHALON			
nucl. sens. n. V			
nucl. principalis n. V	n2, p1(S)	n2(M)	n2, f3
nucl. oralis	n2, p1(S)	n2, p1(S)	n2, f3
nucl. interpolaris			
dorsomedial part	n2	n3, p2	n2, f2
remaining parts	n2	n1, p1	n2, f2

Table 1 (continued)

regions	AChE	SDH	MAO
nucl. sens. n. V			
nucl. caudalis	n3, p1	n2, p1(S)	n2, f2
nucl. descendens lateralis n. V	***	***	***
nucl. commissurae cornuae dorsalis	n2, p3	n2	n1−2, f2
nucl. funiculi dorsalis	n2	n2, p1(S)	n1−2
nucl. vestibulares	p3	n1−2, p3	f3
nucl. cochlearis angularis	n3, p1	n3, p3	n2, f2
nucl. cochlearis magnocellularis	n0−1, p2	n2, p3	n2, f1
nucl. olivaris superior	n1, p2	n3, p2	n2, f2
nucl. solitarius			
rostral part	n1, p1	n2, p2	n2, f2
caudal part	n2, p1	n1,	n2, f2
nucl. parvocellularis medialis	n1, p1	n1, p1(S)	f1
nucl. motorius n. V			
dorsal division	n3, p3	n2, p2	n1, f2
ventral division	n3, p3	n1, p1	n1, f2
nucl. motorius n. VII	n0, p3	n1, p2	n0, f2
nucl. motorius n. X	n2, p3	n1, p0	n0, f2
nucl. n. III	n0, p2	n1−2, p3	n2, f2
nucl. n. IV	n0, p2	n2, p3	n1, f2
nucl. n. XII	n0−1, p3	n2, p3	f2
nucl. reticularis superior	p3(S)	n1, p1	f2
nucl. reticularis inferior	p3(L)	n1, p3(?)	f2
nucl. reticularis lateralis	p3	n1, p1	f2
nucl. reticularis medius	p3	n1, p2	f2
nucl. reticularis colaris	***	***	***
nucl. raphes dorsalis	n2, p2	n2, p1	n2
nucl. tegmentalis dorsalis	n2, p2	n1, p1	n1
nucl. centralis superior	n1, p3	n1, p1	n1−2,
nucl. dorsalis myelencephali	p3	n1, p3	f2
locus coeruleus	n2−3, p3	n1	n2
central gray (rhombencephalon)	n1	n1−2	n2

CEREBELLUM

molecular layer	n3, p1−2	n3	n3, f1
Purkinje cells	0	p2	0
granular layer	n2, p2	n2****	f2
cerebellar nuclei	p1	n1, p2	n1

FIBERS

fibrae cochleares efferentes	f1	f3	f2
tractus mesencephalicus n. V	f1	0	f1
tractus descendens n. V	f1	0	0
fasciculus longitudinalis medialis	f0	f1(S)	0
tractus retroflexus	f0	f3(core)	f1(pere)
tractus interpedunculotegmentalis	f2	f2	f2
root of n. III	f2	f1	f1
root of n. IV	f2	f1	f1
root of n. V	f1	0	0
root of n. VI	f1	f1	f1
root of n. VII	f2	0	f1
root of n. VIII	0	f1(S)	f1
roots of n. IX, X	0	f1	f1
root of n. XII	f2	f1	f1

notes: *, Some activity was found in narrow strips between small tight fasciculi of fibers of the optic tract. **, A thin layer close to the str. fib. gris. sup. (deep) showed stronger activity. ***, This structure does not exist in *Elaphe*. ****, This neuropil may be a glomerulus.

Table 2. AChE, SDH, and MAO activity in *Agkistrodon*. Abbrevations and notes as in Table 1.

regions	AChE	SDH	MAO
MESENCEPHALON			
nucl. lentiformis mesencephali	n2, p1(S)	n1, p3(L)	n2, f1
nucl. geniculatus pretectalis	n3, p1(S)	n3	n1−2, f1
nucl. posterodorsalis	n2	n1	n1
nucl. pretectalis	n1, p1−2(S)	n2−3	n1, f1
paratorus	0	n3, p3	n2, f1
nucl. lat. profundus mesencephali	n1	n1, p2(S)	f1
nucl. profundus mesencephali	p1(S)	n1, p3	n0−1
nucl. opticus basalis	n3, p1	n3, p3	n2
nucl. posterodorsalis tegmentalis	n3, p3	n1, p2	n1
nucl. ruber	n1	n1, p1	n1
nucl. isthmi	n2, p2	n1, p1	n1−2
nucl. of fasc. longitudinalis med.	0	0	n1(?)
nucl. interpeduncularis			
anterolateral part	p1	n2−3	n1−2
medial part (rostral area)	n3, p1	n3	n1−2
medial part (caudal area)	0	n3	n1−2
neuropil of nucl. interpeduncularis	n2(RM)	n3	n3
nucl. tr. mesencephalicus n. V	p3	p3	0
central gray	0	n1−2	n2
optic tectum			
str. zonale	n1	n1−2	n1
str. opticum	0*	0*	f2
str. fib. gris. sup. (superf.)	n1	n2	n1−2, f1
str. fib. gris. sup. (deep)	n2	n2	n1−2, f1
str. griseum centrale			
7b of Kass (?)	n2, p1	n1	n1, f1
str. fibrosum centrale			
7a of Kass (?)	n1, p1(S)	n1, p1(S)	n1, f1
str. griseum periventriculare	n1	n2	n2
str. fibrosum periventriculare	0	n2	n2
str. ependymale	0	0	0−1
RHOMBENCEPHALON			
nucl. sens. n. V			
nucl. principalis	n2	n2, p1(S)	n1, f2
nucl. oralis	n2	n1, p1(S)	n1, f2
nucl. interpolaris			
dorsomedial part	n2−3	n3, p2	f2
remaining part	n2	n1, p1	f2
nucl. caudalis	n1−2(R)	n1−2, p1	f2
nucl. descendens lateralis n. V			
main neuropil	n2	n3, p2	n2, f2
marginal neuropil	n2	n1, p2	n2
nucl. commissurae cornuae dorsalis	n3, p1	n2−3, p1	n1
nucl. funiculi dorsalis	n3	0	0
nucl. vestibulares	p1−2	n1, p3	n1−2
nucl. cochlearis angularis	n2	n3, p3	n1−2
nucl. cochlearis magnocellularis	n1	n2, p3	n1
nucl. olivaris superior	0	n3, p1−2	f2
nucl. solitarius			
rostral part	n1	n2, p2	n2
caudal part	n2	n1	n2
nucl. parvocellularis medialis	n1, p1	n1, p1	0
nucl. motorius n. V			
dorsal division	n2, p2	n1, p2	f1
ventral division	n2, p2	n1, p0−1	f1
nucl. motorius n. VII	n1, p2	n1, p1−2	0
nucl. motorius n. X	n2, p1−2	n1	n1
nucl. n. III	n1, p1	n2, p2−3	n1
nucl. n. IV	n2, p2	n2, p2−3	n1

Table 2 (continued)

regions	AChE	SDH	MAO
nucl. n. VI	n1, p1	n1, p2−3	n0−1
nucl. n. XII	p1−2	n1, p2−3	0
nucl. reticularis superior	p3(S)	p1	0
nucl. reticularis inferior	p1(L)	p1	0
nucl. reticularis lateralis	0	n2, p3	n1
nucl. reticularis medius	p1(S)	p1	0
nucl. reticularis caloris	n1, p1	n3, p2−3	f1
nucl. raphe dorsalis	n3, p3	n2, p2	n2
nucl. tegmentalis dorsalis	n2, p2	n2, p1	n1
nucl. centralis superior	p0−1	n1, p1−2	n1−2
nucl. dorsalis myelencephali	n0−1, p1	n1, p2	0
locus coeruleus	n2, p2	n1	n1−2
central gray (rhombencephalon)	n1,	n1−2	n2
CEREBELLUM			
molecular layer	n3	n3	n3
Purkinje cells	p1	p1−2	0
granular layer	n1, p2−3	n2−3****	f2
cerebellar nuclei	0	n1, p3	f1
FIBERS			
fibrae cochleares efferentes	f1	f3	f3
tractus mesencephalicus n. V	f2	f1	0
tractus descendens n. V	f2(per)	f1	f1(per)
fasciculus longitudinalis medialis	f2(S)	f1(S)	0
tractus retroflexus	0	f3(core)	f1(per)
tractus interpedunculotegmentalis	f3	f1(?)	f2
root of n. III	f2	f1(?)	f1
root of n. IV	f2	f1	f1(?)
root of n. V	f2	f0−1	f1
root of n. VI	f1	f1	f1
root of n. VII	f2−3	f1(?)	f1
root of n. VIII	f1	f1	f1
roots of n. IX, X	f1	f0−1	f1
root of n. XII	f2	f1	f1

superficial layer in both species. In *Elaphe* the deep sublayer showed strong AChE activity, and in *Agkistrodon* moderate AChE activity. In the sgc AChE activity was strong in both species, but it was stronger in *Elaphe*. In *Agkistrodon* the entire sfgs was thinner than in *Elaphe*; the AChE activity in its deep sublayer had about the same strength as in the sgc. In *Elaphe*, since the intensity of activity was stronger, the stratification of the TO appeared clearer than in *Agkistrodon* (Figs. 1-B, C, D; 2-A, B, C, D; 3-B; 5-B; 6-A; 7-A; and 9).

The sfgs did not show 2 layers in either the SDH or MAO staining, but the activity of these 2 enzymes was clearly demonstrated throughout the layer. Numerous fibers within the TO, especially in the stratum opticum of *Agkistrodon* and the sfc of both species, also showed MAO activity.

Cells of the nucl. tr. mesencephalicus n. trigemini, which were found mainly in the str. fibrosum periventriculare (sfp) and the str. griseum periventricu-

lare (sgp), showed marked SDH and AChE activity (Figs. 1-C, D; 2-D; 4; and 6-A). They were crowded into a group at the caudomedian portion of the TO in both species.

Tegmentum of the midbrain

The nucl. profundus mesencephali of SCHROEDER (1981 a) (pm) is situated in a location similar to that of the nucl. Z of HUBER and CROSBY (1926), and is a prominent structure at the ventrolateral surface of the brainstem in snakes. It showed strong AChE activity in the perikarya in *Elaphe* (Fig. 1-B), but only weak activity in *Agkistrodon*. As the caudal end of the pm approached the surface of the mesencephalic tegmentum, the nucl. opticus basalis appeared ventromedially to it and showed moderate to intense AChE and SDH activity in the perikarya in *Elaphe*, and moderate to intense SDH activity only in the perikarya in *Agkistrodon*. In both species there was intense

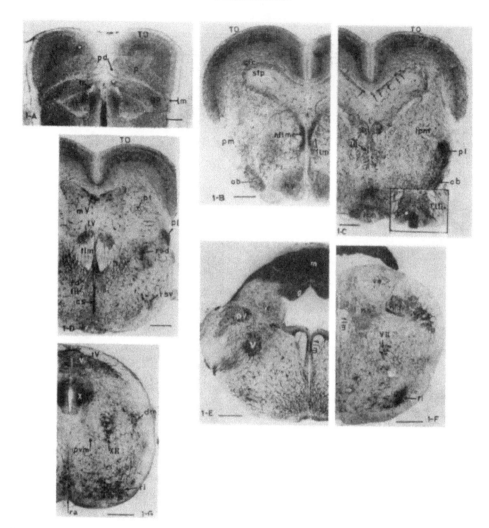

Figs. 1-A to G. A series of transverse sections though the brain of *Elaphe*, illustrated from rostral (A) to caudal (G), AChE. Bar represents 500 µm. Capillaries react for AChE in the preparations.

A: The rostral part of the TO. Arrow pd indicates the nucl. posterodorsalis, which lacks activity. The tr. retroflexus (tr) also shows lack of activity.

B: The caudal to Fig. 1-A. Laminations are observed in the TO. The nucl. of the fasciculus longitudinalis medialis (nflm) shows prominent activity.

C: The level of the oculomotor nucleus. The nucl. mesencephalicus n. trigemini is indicated by arrows. Subdivisions of the nucl. interpeduncularis are demonstrated by grade of activity. Inset is the rostral part of the nucl. interpeduncularis.

D: The caudal part of the TO. The nuclei within the reticular formation and raphe show more activity than in *Agkistrodon*.

E: The level of the cerebellum.

F: The level of the nucl. interpolaris of the nucl. descendens n. V. (iVm, iV).

G: The level of the caudal part of the rhombencephalon.

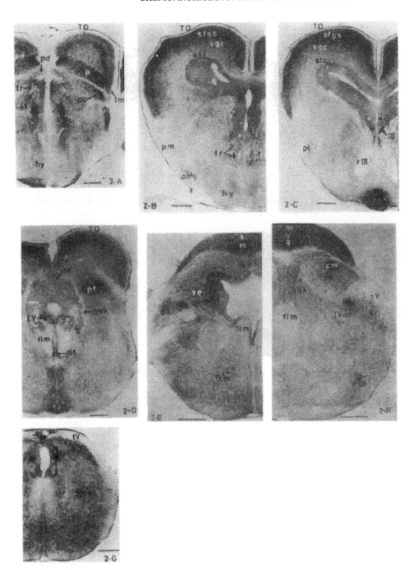

Figs. 2-A to G. A series of transverse sections through the brain of *Elaphe*, illustrated from rostral (A) to caudal (B). SDH. Bar represents 500 μm.

A: The level of the rostral part of the TO.

B: The level caudal to Fig. 2-A.

C: The level of the nucl. n. III.

D: The caudal part of the TO.

E: The level of the cerebellum. Scattered Purkinje cell bodies showing activity can be seen at higher magnification in the molecular layer of the cerebellum.

F: The level of the nucl. interpolaris of the nucl. descendens n. V. The pars medialis of the nucl. interpolaris (iVm) is distinguished by activity. In comparison with AChE preparations, no strongly active cells are found in the reticular formation.

G: The caudal part of the rhombencephalon.

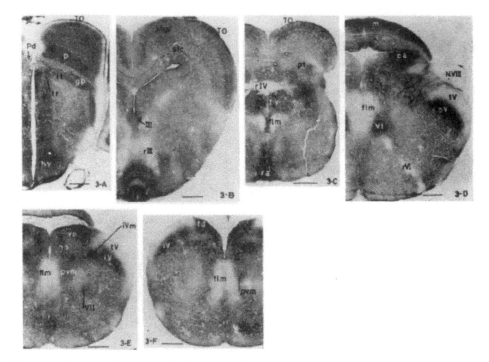

Figs. 3-A to F. A series of transverse sections through the brain of *Elaphe*, illustrated from rostral (A) to caudal (F) MAO. Bar represents 500 μm.

A: The level of the commissura posterior. Most fibers, which are included in the tr. retroflexus (tr) reacted for MAO.
B: The level of the oculomotor nucleus. The nucl. interpeduncularis (ip) shows prominent activity as in other vertebrates.
C: The level of the caudal part of the TO.
D: The level of the root of the nucl. n. VIII.
E: The middle part of the rhombencephalon.
F: The level caudal to Fig. 3-A.

AChE activity in the nucl. posterolateralis tegmentalis of SCHROEDER (1981b) (Figs. 1-C and 5-B). This nucleus was substituted for the pm at the caudolateral part of the mesencephalic tegmentum and caudodorsal to the nucl. opticus basalis, and it could be traced caudally to the level of the nucl. isthmi pars parvocellularis of MOLENAAR (1977).

In *Elaphe* perikarya with marked AChE activity could be seen along the fasciculus longitudinalis medialis, medially to its fasciculus, from the rostral tip of the nucl. oculomotorius to the junction between the mesencephalon and the diencephalon (Figs. 1-B and 4), but not in *Agkistrodon*. This neuronal group appeared to be the nucl. of the fasciculus longitudinalis medialis described in the turtle by TUGE (1932). Activity of SDH and MAO in this nucleus was indistinguishable from that in the surrounding neural tissue.

The paratorus (SENN and NORTHCUTT, 1973), which is peculiar to snakes, showed prominent SDH activity in cells and neuropil, and appeared between fibers of the tr. tectobulbaris at the caudal level of the TO (Figs. 2-D). The nucl. interpeduncularis of both species, generally speaking, stained strongly for SDH, AChE, and MAO, but there were variable reactions among its subdivisions, as shown in Tables 1 and 2 (Figs. 1-C; 2-C; 3-B; 4; 5-B; and 6-A). The nucl. ruber, which is located in the formatio reticularis lateral to the root of the n. III, showed stronger AChE activity in *Elaphe* than in *Agkistrodon*. The nucl. isthmi showed moderate AChE and MAO activity, and weak SDH activity.

Medulla oblongata

In the nucl. tr. descendens n. trigemini (dV), there was a round limited position at the dorsomedial part

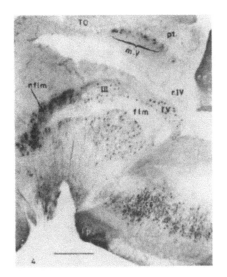

Fig. 4. Paramedian sagittal sections of the brain of *Elaphe*, AChE. Bar represents 500 μm. The nucl. of the fasciculus longitudinalis medialis (nflm) and the nucl. centralis superior (cs) are strongly demonstrated.

The most peculiar structure of the trigeminal nucleus of *Agkistrodon* was the nucl. of the lateral descending trigeminal tract (dlV) of MOLENAAR (1974). It had prominent SDH activity in the main neuropil, weak activity in the marginal neuropil, and moderate activity in the perikarya (Figs. 6-C and 8-B). There was moderate AChE and MAO activity in both the main and marginal neuropil (Figs. 5-D and 8A). There were also many capillaries showing AChE reaction in the dlV (Figs. 5-D and 8-A). There is a particular nucleus in the formatio reticularis which is found exclusively in the Crotalinae (the "new nucleus" of KISHIDA et al., 1980 or the "nucl. reticularis caloris" of GRUBERG et al., 1979), which is located in the ventrolateral reticular formation, ventral to the nucl. descendens n. V at the level between the nucl. oralis and the nucl. interpolaris. There was high SDH activity in the perikarya of this specific nucleus, but weak AChE activity. In its neuropil there was negligible to weak SDH activity, and negligible activity of MAO (Figs. 5-C and 6-B). As in the dlV, strong AChE activity was seen in the capillary walls. Other nuclei of the reticular formation showed intense AChE activity in the perikarya in *Elaphe* (Figs. 1-D, E, F and G), but only weak activity in *Agkistrodon* (Figs. 5-C, D). SDH activity was weak in these nuclei in both species, but was relatively strong in the nucl. reticularis lateralis of the lower brainstem in *Agkistrodon*.

of the nucl. interpolaris which was stained strongly for SDH, but the rest of the nucleus was only weakly to moderately positive (Figs. 1-F and 2-F). The same position did not stand out very strongly in AChE and MAO staining.

The nucleus in our material that we interpreted as the nucl. centralis superior extended rostrally from the rostral raphe to the level that included the caudal

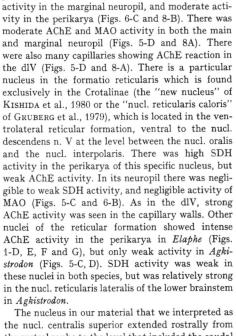

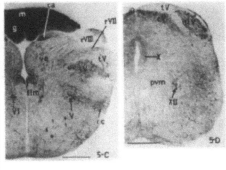

Figs. 5-A to D. Transverse sections of the brain of *Agkistrodon*, AChE. Bar represents 500 μm.

A: Section through the rostral part of the TO.

B: The level of the middle part of the TO. The laminations of the TO are indistinct in comparison with Figs. 1-B, C, and D.

C: The level of the root of the n. VIII. In the nucl. reticularis caloris (rc), considerable activity is found in the walls of capillaries, which are seen as small dots.

D: The caudal level of the medulla oblongata. The dlV is located at the dorsolateral border of the medulla.

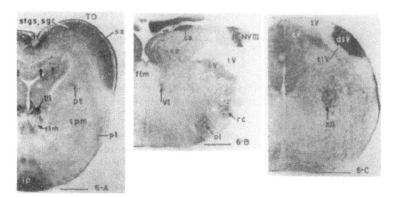

Figs. 6-A to C. Transverse sections of the brain of *Agkistrodon*, SDH. Bar represents 500 μm.

A: The level of the nucl. interpeduncularis. The arrow indicates the nucl. tr. mesencephali n. V.
B: The level of the root of the n. VIII. The nucl. reticularis caloris (rc) is clearly demonstrated
C: The caudal part of the rhombencephalon. The strongest activity is found in the dIV.

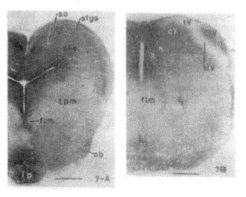

Figs. 7-A and B. Transverse sections of the brain of *Agkistrodon*, MAO. Bar represents 500 μm.

A: The middle part of the TO.
B: The level of the dIV. Considerable activity is found in the dIV.

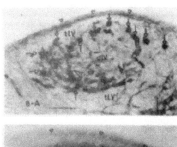

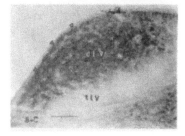

interpeduncular nucleus and the nucl. n. IV. This nucleus consisted of small and medium-sized cells which were crowded vertically along the raphe, and topologically resembled the nucl. centralis superior of TABER et al. (1960) in the cat. Perikarya in the nucleus showed strong AChE activity in *Elaphe* (Figs. 1-D and 4), but negligible or weak activity in *Agkistrodon*.

Cerebellum (Figs. 1-E, F; 2-E, F; 3-D; and 5-C)

In *Agkistrodon*, the cerebellum consisted of 3 layers from the outer surface to the IVth ventricle: a mole-

Figs. 8-A to C. Transverse sections of the dIV of *Agkistrodon*. Bar represents 100 μm. A: AChE; B: SDH; C: MAO. Open triangles indicate the marginal neuropil of the dIV, and arrows indicate capillaries.

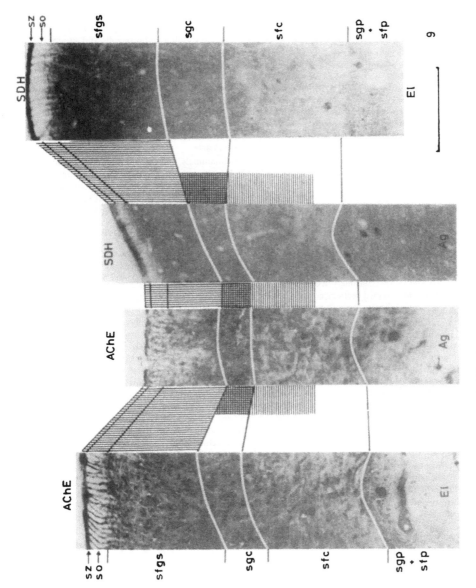

Fig. 9. Transverse sections of the TO of *Elaphe* and *Agkistrodon*.
Bar represents 250 μm. Ag. *Agkistrodon*; El. *Elaphe*. ▦, area receiving visual input; ▤, area receiving infrared input.

cular layer, a Purkinje cell layer, and a granular layer. The Purkinje layer was thick, consisting of several layers of Purkinje cells which showed weak AChE activity. This layer was not present in *Elaphe*, but Purkinje cells showing negligible AChE activity were scattered in the molecular layer together with some other small cells having weak to moderate activity. In both species the neuropil of the molecular layer showed strong activity of all 3 enzymes. In the granular layer of both species, SDH activity was seen in what we interpreted as glomeruli, especially in the mid-caudal portion of the layer, and was also scattered sporadically in some perikarya. AChE was seen in both perikarya and neuropil in this layer in both species.

Fibers

The fibers at the roots of the cranial nerves of both species, as a rule, exhibited weak or moderate activity of all 3 enzymes, but the root of *n*. VII in *Agkistrodon* was stained rather heavily for AChE. There was high activity of SDH in the core of the tr. retroflexus, and weak MAO activity in fibers in its peripheral portion (Figs. 2-B and 3-A). A strongly AChE-positive zone, which in Bodian-stained sections could be seen to be composed of fine fibers, arose from the caudodorsal part of the nucl. interpeduncularis and extended along the raphe toward the area between the fasciculus longitudinalis medialis of both sides at the level of the nucl. n. IV, passing between a pair of small cell groups. Topologically, these cell groups seemed to correspond to the nucl. raphe dorsalis of mammals. The zone could be followed to the AChE-positive ventrolateral part of the periventricular gray of the caudal end of the mesencephalon via the undersurface or along the caudal end of the nucl. n. IV. This ventrolateral part contained oval and round small cells stained lightly by Nissl staining. For this cell group we have provisionally adopted the name nucl. tegmentalis dorsalis of the snake, since it appeared to correspond to the nucleus of that name in mammals. The zone also had a few connections to the nucl. n. IV via fine AChE-positive fibers. For convenience of description we have named the bundle the tr. interpedunculotegmentalis of the snake (Fig. 1-D).

Discussion

We have demonstrated obvious structural and histochemical differences in the brainstem of *Agkistrodon*, which possesses pit organs, and *Elaphe*, which does not. Although the basic distribution of activity was similar in both species, differences could be observed especially in the TO, in the relay nuclei of the infrared sensory system, in the nucleus of the fasciculus longitudinalis medialis, and in the nucl. centralis superior.

As briefly stated in a previous paper (KUSUNOKI et al., 1971), as a general rule SDH activity is found mainly in somatic regions, AChE activity is seen in both visceral and somatic regions and in areas with correlative function, and MAO activity is seen mainly in visceral and limbic regions. In the present study on the snake brain, the pattern of SDH and AChE activity was similar to that seen in other vertebrates. MAO, instead, was present throughout the somatic neuropil, as well as in the visceral neuropil, and in most fibers in the brainstem.

Visual system

In nonmammals, the TO is the primary visual center, which integrates information from the external environment as one of the animal's important correlation centers. In the TO, prominent AChE activity has been demonstrated in areas receiving retinal projections in the lamprey (WÄCHTLER, 1974), bony fishes (WAWRZYNIAK, 1962; ISHIBASHI, 1965; CONTESTABILE and ZANNONI, 1975; KISHIDA et al., 1976; VILLANI et al., 1979), amphibians (SHEN et al., 1955; GRUBERG and GREENHOUSE, 1973; CIANI and FRANCESCHINI, 1982), reptiles (MINELLI, 1970; KUSUNOKI, 1971; SETHI and TEWARI, 1976; CONTESTABILE and CUPPINI, 1977; VILLANI and CONTESTABILE, 1982), and birds (KUSUNOKI, 1969/70; MINELLI, 1970; CIANI et al., 1978). Very low or negligible activity has been demonstrated in the shark (KUSUNOKI et al., 1973) and in the hagfish (WÄCHTLER, 1975; KUSUNOKI et al., 1982).

In *Elaphe*, strong or moderate SDH and AChE activity also appeared in superficial strata, i.e., in the str. zonale, in the sfgs, and in the str. opticum between small tight fasciculi of fibers of the optic tract. These strata are innervated by fibers from the retina (HALPERN and FRUMIN, 1973; NORTHCUTT and BUTLER, 1974; SCHROEDER, 1981a). It is not clear why the sfgs was divided into 2 sublayers by AChE staining.

Using electrophysiological techniques, KASS et al. (1978) recorded both visual and infrared responses from layer 7b, which corresponds to our sgc, of the TO of rattlesnakes (pit vipers like the *Agkistrodon* we used). SCHROEDER (1981a) found retinal projections in the upper part of the sgc in rattlesnakes. This layer showed stronger AChE activity than other layers of the TO in *Agkistrodon* in our work, and this may reflect the bimodal physiological activity of this layer.

MASAI (1973) reported that the sfgs was thicker in diurnal snakes than in nocturnal snakes, and in our work also the sfgs was thicker in *Elaphe* than in *Agkistrodon*, which may reflect the diurnal habits of *Elaphe* and the more nocturnal habits of *Agkistrodon*.

In the TO in both species, MAO reaction was diffuse in neuropil and fibers. In the turtle, the superficial layers show weak MAO activity and the deep layers corresponding to the sgp and sfp stain moderately (KUSUNOKI, 1971). Monoamine terminals have been demonstrated in the sfgs, sfp, sfc, and so of the turtle by histofluorescent and immunohistochemical techniques (PARENT, 1973; UEDA et al., 1983).

Other areas receiving retinal projections in the pretectum and mesencephalon have been investigated experimentally in snakes which do not possess pit organs, i.e., in *Natrix* (ARMSTRONG, 1951), in *Thamnophis sirtalis* (HALPERN and FRUMIN, 1973), in *Natrix sipedon sipedon* (NORTHCUTT and BUTLER, 1974, and *Vipera aspis* (REPÉRANT and RIO, 1976), as well as in pit vipers (SCHROEDER, 1981a). In our histochemical preparations, a clear difference was seen in the nucl. posterodorsalis, which showed moderate AChE activity and weak SDH and MAO activity in the neuropil of *Agkistrodon*, but no AChE activity, weak SDH activity, and moderate MAO activity in *Elaphe*. Enzymatic distribution similar to that of *Elaphe* is found in the area pretectalis of the turtle, which corresponds to the nucl. posterodorsalis of the snake, in which activity of AChE and SDH is lacking (KUSUNOKI, 1971). There was also difference in density of SDH activity in the nucl. geniculatus pretectalis. In other nuclei receiving retinal projections, including the pretectal nuclei and the nucl. opticus basalis, the results were variable, as shown in Tables 1 and 2. The reason for this variability is not clear.

Infrared system

Impulses originating from the pit organs are conveyed to the dlV (MOLENAAR, 1974, 1977, 1978a; SCHROEDER and LOOP, 1976). In the python, secondary fibers from the dlV project directly to the contralateral sgc of the TO (MOLENAAR and FIZAN-OOSTVEEN, 1980; NEWMAN et al., 1980), but in *Crotalus* they project to the nucl. reticularis caloris between the TO and dlV (GRUBERG et al., 1979; KISHIDA et al., 1980; NEWMAN et al., 1980; STANFORD et al., 1981). These nuclei with somatic functions exhibited strong to moderate SDH activity in the neuropil and/or perikariya. It is interesting that the dlV, which participates in this particular sensation, was fed by many capillaries, and showed remarkable activity of SDH, which is concerned with aerobic metabolism in the neuropil and perikarya. In the present study weak reaction for AChE in the perikarya was found in these nuclei, but NEWMAN et al. (1981) have reported marked AChE activity in cells of these nuclei in the rattlesnake and the python. This apparent contradiction might be due to species variation caused by their natural history and suggests that the mechanism of neurotransmission in

the infrared system may differ according to the species; or it may be due to difference in the methods used to detect AChE activity. Further work is needed to make these points clear.

Although MAO activity was detectable in the neuropil and fibers within the dlV, the serotonin-containing fibers were much finer than those of the common sensory trigeminal system (KADOTA et al., 1984). The marginal neuropil, which receives vagal afferents (KISHIDA et al., 1984), was distinguished from the main neuropil by the intensity of SDH activity in the somatic areas. This phenomenon might indicate a functional difference in these neuropils, i.e. the main neuropil has a somatic function, and the marginal neuropil has a visceral one.

Common sensory trigeminal system

The nuclei of the trigeminal system of *Elaphe* and *Agkistrodon* have a construction similar to that of the python reported by MOLENAAR (1977, 1978a). The distribution pattern of enzymatic activity was essentially alike in snakes studied. As stated in Results, a clearly delineated round area was found at the dorsomedial part of the nucl. interpolaris n.V, which was distinguished from its surroundings by the presence of strong SDH activity, and which also showed MAO activity. This round portion receives fibers of the V³ nerves in snakes (MOLENAAR, 1978b), in birds (ARENDS, 1981), and in mammals (e.g. ASTRÖM, 1953, TORVIK, 1956). It also receives fibers of the IX and X nerves in reptiles (Kishida et al., in preparation), in birds (DUBBELDAM, 1979), and in mammals (BECKSTEAD and NORGEN, 1979). Therefore, the round portion might play some special role in the sensory trigeminal system.

Nucl. profundus mesencephali of SCHROEDER

The nucl. profundus mesencephali of SCHROEDER is a prominent structure in the snake brain which receives no retinal projections (HALPERN, 1973; HALPERN and FURMIN, 1973; NORTHCUTT and BUTLER, 1974; REPÉRANT and RIO, 1976; SCHROEDER, 1981a). SCHROEDER (1981a, b) remarked that the nucl. profundus mesencephali may be homologous to the nucl. rotundus of other reptiles because it receives heavy tectal projections, but this question remains to be investigated. A recent experimental study on the nonretinotopic projections in snakes has identified the nucl. rotundus in a part of the nucl. lentiformis thalami of Northcutt and Butler (DACEY and ULINSKI, 1983). In our histochemical study, it was implied that the somatic function of the nucleus was modified according to the species, since the patterns of enzymatic activity differed. The reason for this variation remains in obscure.

Nucl. posterolateralis and dorsolateralis tegmentalis of SCHROEDER

Although it seems that the nucl. posterolateralis tegmentalis is comparable with the substantia nigra of HUBER and CROSBY (1933) at the level caudal to the nucl. opticus basalis, SCHROEDER (1981a) thought it possible that this nucleus resembles the nonmammalian nucl. isthmi, on the basis of fiber connections between this nucleus, which shows heavy AChE activity, and the TO. Also, we have confirmed the AChE activity in the neuropil and perikarya of the nucleus. NEWMAN et al. (1980) found heavy activity of AChE in the dorsolateral posterior tegmental nucleus in the python and rattlesnake, but no activity was demonstrated in our AChE preparations.

Nucl. of the fasciculus longitudinalis medialis, nucl. centralis superior, raphe nuclei, and reticular nuclei

As shown in the Results, marked differences between *Agkistrodon* and *Elaphe* were demonstrated in these nuclei, in which the perikarya showed strong AChE activity in *Elaphe*, but not in *Agkistrodon*. *Agkistrodon* also showed negligible activity of SDH in the nucl. of the fasciculus longitudinalis medialis and the raphe nuclei. Further work based on ecological considerations is needed in these nuclei.

Nucl. interpeduncularis and related tracts

No prominent difference was found in the two species, as a rule. SDH and MAO activity was demonstrated strongly in all subdivisions of the nucleus, and marked AChE activity was exhibited mainly in the rostromedial portion. The nucl. interpeduncularis, which is phylogenetically old region, was stained intensely for SDH and MAO in all vertebrates studied, as stated in a previous report of KUSUNOKI (1971). As regards AChE activity in the nucleus, the major groups of vertebrates have prominent activity in the nucleus; for example, the cyclostomes (WÄCHTLER, 1974, 1975; KUSUNOKI, 1982), the teleosts (ISHIBASHI, 1967; KISHIDA et al., 1976), the frogs (SHEN, 1955), the birds (KUSUNOKI, 1969/70), and the mammals (FRIEDE, 1966; RAMON-MOLINER, 1972; PALKOVITS and JACOBOWITZ, 1974; WILSON and WATSON, 1980). Contrary to these animals, only weak activity of AChE is present in the shark (KUSUNOKI, 1973), and the turtle (KUSUNOKI, 1971). Therefore, the snakes in this study belonged to the major group. The tr. retroflexus, which connects the nucl. habenulae and the nucl. interpeduncularis exhibits definite MAO activity in most animals. In snakes, staining difference was seen in the tract, i.e., concentrated SDH activity was found in fibers constituting the core of the tract, but activity of MAO was found in peripherally arranged fibers of the tract. There are species difference in SDH preparations; for instance, the tr. retroflexus of the turtle shows lack of SDH activity (KUSUNOKI, 1971), strong activity in Suncus (insectivora), and weak or no activity in the rat, gerbil, hamster (KADOTA and KUSUNOKI, private observations), and bat (KUSUNOKI and TSUDA, 1968). The tract, which may correspond to the tr. interpedunculotegmentalis of mammals which exhibits prominent activity of AChE (WILSON and WATSON, 1980), was detected in the snake brain as in mammals, and could be traced easily into the ventrolateral part of the periventricular gray by AChE staining.

Finally, our results suggest that *Elaphe*, which has a well differentiated TO, is active diurnally and seizes prey by visual information. On the other hand, *Agkistrodon* has less differentiation in the visual regions of the TO, which suggests that it is less dependent on visual acuity, and therefore probably more active at night, at which time its sensitivity to infrared information strongly complements its visual sense in detecting and capturing prey.

Abbreviations

nucl. n. III	III
nucl. n. IV	IV
nucl. motorius n. V	V
nucl. n. VI	VI
nucl. motorius n. VII	VII
nucl. motorius n. X	X
nucl. n. XII	XII
nucl. caudalis n. V	c. V
nucl. cochlearis angularis	ca
cerebellum	ce
nucl. cochlearis magnocellularis	cm
nucl. centralis superior	cs
nucl. descendens lateralis n. V	dlV
nucl. dorsalis myelencephali	dm
nucl. funiculi dorsalis	fd
fasciculus longitudinalis medialis	flm
granular layer (cerebellum)	g
nucl. geniculatus pretectalis	gp
hypothalamus	hy
nucl. interpolaris (remaining parts)	iV
nucl. interpolaris (dorsomedial part)	iVm
nucl. interpeduncularis	ip
neuropil of the nucl. interpeduncularis	ipn
nucl. lentiformis mesencephali	lm
nucl. lateralis profundus mesencephali	lpm
nucl. lentiformis thalami	lt
molecular layer (cerebellum)	m
nucl. tr. mesencephalicus n. V	mV
nucl. of fasc. longitudinalis medialis	nflm
N. VIII	N. VIII
nucl. solitarius	ns
nucl. solitarius (rostral part)	nsr
nucl. oralis n. V	oV
nucl. opticus basalis	ob
nucl. olivaris superior	ol

nucl. pretectalis	p
nucl. sens. principalis n. V	pV
nucl. posterodorsalis	pd
nucl. posterolateralis tegmentalis	pl
nucl. profundus mesencephali	pm
paratorus	pt
nucl. parvocellularis medialis	pvm
root of n. III	r. III
root of n. IV	r. IV
root of n. VI	r. VI
root of n. VII	r. VII
root of n. VIII	r. VIII
nuclei raphes	ra
nucl. reticularis caloris	rc
nucl. raphe dorsalis	rd
nucl. reticularis inferior	ri
nucl. reticularis medius	rm
nucl. reticularis superior (dorsal part)	rsd
nucl. reticularis superior (ventral part)	rsv
stratum fibrosum centrale	sfc
stratum fibrosum et griseum superficiale	sfgs
stratum fibrosum periventriculare	sfp
stratum griseum centrale	sgc
stratum griseum periventriculare	sgp
stratum opticum	so
stratum zonale	sz
optic tectum	TO
tr. descendens n. V	tV
tr. interpedunculotegmentalis	tit
tr. descendens lateralis n. V	tlV
tr. mesencephalicus n. V	tmV
tr. retroflexus	tr
N. mandibularis	V²
nuclei vestibulares	ve

Literature

ARENDS, J. J. A.: Sensory and motor aspects of the trigeminal system in the mallard (*Anas platyrhynchos* L.): A contribution to the biocybernetics of feeding. Thesis of Unversity of Leiden, 1–167 (1981).

ARMSTRONG, J. A.: An experimental study of the visual pathways in a snake (*Natrix natrix*). J. Anat., 85, 275–288 (1951).

ASTRÖM, K. E.: On the central course of afferent fibers in the trigeminal, facial, glossopharyngeal, and vagal nerves and their nuclei in the mouse. Acta Physiol. Scand. 29 Suppl., 209–320 (1953).

BECKSTEAD, R. M., and NORGEN, R.: An autoradiographic examination of the central distribution of the trigeminal, facial, glossopharyngeal, and vagal nerves in the monkey. J. Comp. Neurol., 184, 455–472 (1979).

BRAND, S. and MUGNANI, E.: Pattern of distribution of acetylcholinesterase in the cerebellar cortex of pond turtle, with emphasis on parallel fibers. A histochemical and biochemical study. Anat. Embryol., 158, 271–287 (1980).

CIANI, F., CONTESTABILE, A., and VILLANI, L.: Acetylcholinesterase activity in the normal and retino-deprived optic tectum of the quail. Light and electron microscopic histochemistry and biochemical determination. Histochem., 59, 81–95 (1978).

CIANI, F., and FRANCESCHINI, V.: the pattern of acetylcholinesterase distribution in the normal and retino-deprived optic tectum of newt. Z. mikrosk.-anat. Forsch., 96, 600–612 (1982).

CONTESTABILE, A.: Histochemical localization of acetylcholinesterase in the cerebellum of some seawater teleosts. Brain Research, 99, 425–429 (1975).

CONTESTABILE, A. and ZANNONI, N.: Histochemical localization of acetylcholinesterase in the cerebellum and optic tectum of four freshwater teleosts. Histochem., 45, 279–288 (1975).

CONTESTABILE, A. and CUPPI, S.: Enzymatic patterns in reptilian brain. Histochemical characterization of the optic tectum. Experientia, 33, 757–759 (1977).

DACEY, D. M. and ULINSKI, P. S.: Nucleus rotundus in a snake, *Thamnophis sirtalis*: An Analysis of a nonretinotopic projection. J. Comp. Neurol., 216, 175–191 (1983).

DE COCK BUNING, Tj., GORIS, R. C., and TERASHIMA, S.: The role of thermosensitivity in the feeding behavior of the pit viper *Agkistrodon blomhoffi brevicaudus*. Jap. J. Herpet., 9, 7–27 (1981).

DE COCK BUNING, Tj.: Thermal sensitivity as a specialization for prey capture and feeding in snakes. Amer. Zool., 23, 363–375 (1983).

DUBBELDAM, J. L., BRUS, E. R., MENKEN, S. B. J., and ZEILSTRA, S.: The central projections of the glossopharyngeal and vagus ganglia in the mallard, *Anas platyrhynchos* L., J. Comp. Neurol., 183, 149–168 (1979).

FRIEDE, R. L.: Topographic brain chemistry. 241–281, Academic Press, New York, London (1966).

GLENNER, G., BURTNER, H. J., and BROWN, G. W.: The histochemical demonstration of monoamine oxidase activity by tetrazolium salts. J. Histochem. Cytochem., 5, 591–600 (1957).

GRUBERG, E. R., and GREENHOUSE, G. A.: The relationship of acetylcholinesterase activity to optic fiber projections in the tiger salamamder *Ambystoma tigrinum*. J. Morph., 141, 147–156 (1973).

GRUBERG, E. R., KICLITER, E., NEWMAN, E. A., KASS, L., and HARTLINE, P. H.: Connections of the tectum of the rattlesnake *Crotalus viridis*: an HRP study. J. Comp. Neurol., 188, 31–42 (1979).

HALPERN, M.: Retinal projections in blind snakes. Science, 182, 390–391 (1973).

HALPERN, M., and FRUMIN, N.: Retinal projection in a snake, *Thamnophis sirtalis*. J. Morph., 141, 359–382 (1973).

HUBER, G. C. and CROSBY, E. C.: On thalamic and tectal nuclei and fiber paths in the brain of the american alligator. J. Comp. Neurol., 40, 97–227 (1926).

HUBER, G. C. and CROSBY, E. C.: The reptilian optic tectum. J. Comp. Neurol., 57, 57–163 (1933).

ISHIBASHI, H.: Comparative neurological studies on the acetylcholinesterase distribution in the central nervous system of the goldfish. Acta Anat. Nippon., 40, 5, Supp. p. 4 (1965). (in Japanese).

KADOTA, T., KISHIDA, R., GORIS, R. C., KUSUNOKI, T., TERASHIMA, S., and YAMADA, H.: Serotonin-immunoreactive fibers in the trigeminal sensory nuclei of snakes with infrared sensitivity. Neurosc. Letter, Supp. 17, S 117 (1984).

KASS, L., LOOP, M. S., and HARTLINE, P. H.: Anatomical and physiological localization of visual and infrared cell layers in tectum of pit vipers. J. Comp. Neurol., 182, 811 to 820 (1978).

KISHIDA, R., KIRIYAMA, H., and KUSUNOKI, T.: The chemoarchitectonics of the teleostean brain: *Gobiidae*. Acta Anat. Nippon., 51, 4, 3 (1976).

KISHIDA, R., AMEMIYA, F., KUSUNOKI, T., and TERASHIMA, S.: A new tectal afferent nuclei of the infrared sensory system in the medulla oblongata of crotaline snakes. Brain Research, 195, 271–279 (1980).

KISHIDA, R., YOSHIMOTO, M., KUSUNOKI, T., GORIS, R. C., and TERASHIMA, S.: Vagal afferent C fibers projecting to the lateral descending trigeminal complex of crotaline snakes. Exp. Brain Res., 53, 315 – 319 (1984).

KOELLE, G. B., and FRIEDENWARLD, J. S.: A histochemical method for localizing cholinesterase activity. Proc. Soc. Exp. Biol. Med., 70, 617 – 622 (1948).

KUSUNOKI, T., and TSUDA, Y.: Comparative neurological studies on the distribution of some enzymes in the central nervous system of the bat. Acta Anat. Nippon., 43, 1, 88 (1968) (in Japanese).

KUSUNOKI, T.: The chemoarchitectonics of the avian brain. J. Hirnforsch., 11, 477 – 497 (1969/70).

KUSUNOKI, T.: The chemoarchitectonics of the turtle brain. Yokohama Med. Bull., 22, 1 – 29 (1971).

KUSUNOKI, T., TSUDA, Y., and TAKASHIMA, F.: The chemoarchitectonics of the shark brain. J. Hirnforsch., 14, 13 – 26 (1973).

KUSUNOKI, T., KADOTA, T., and KISHIDA, R.: Chemoarchitectonics of the brain stem of the hagfish, *Eptatretus burgeri*, with special reference to the primordial cerebellum. J. Hirnforsch., 23, 109 – 119 (1982).

MASAI, H.: Structure patterns of the optic tectum in Japanese snakes of the family Colubridae, in relation to habit. J. Hirnforsch., 14, 365 – 374 (1973).

MALER, L., COLLINS, M., and MATHIESON, B.: The distribution of acetylcholinesterase and cholineacetyl transferase in the cerebellum and posterior lateral line lobe of weakly electric fish (*Gymnotidae*). Brain Research, 226, 320 – 325 (1981).

MINELLI, G.: Histochemical studies on certain enzymatic activities of the encephalon of *Testudo graeca* and *Coturnix coturnix*. Riv. Biol., 63, 61 – 86 (1970).

MOLENAAR, G. J.: An additional trigeminal system in certain snakes possessing infrared receptors. Brain Research, 78, 340 – 344 (1974).

MOLENAAR, G. J.: The rhombencephalon of *Python reticulatus*, a snake possessing infrared receptors. Netherlands J. Zoology, 27, 133 – 180 (1977).

MOLENAAR, G. J.: The sensory trigeminal system of a snake in the possession of infrared receptors. I. The sensory trigeminal nucleus. J. Comp. Neurol., 179, 123 – 136 (1978a).

MOLENAAR, G. J.: The sensory trigeminal system of a snake in the possession of infrared receptors. II. The central projections of the trigeminal nerve. J. Comp. Neurol., 179, 137 – 152 (1978b).

MOLENAAR, G. J., and FIZAAN-OOSTVEEN, J. L. F. P.: Ascending projections from the lateral descending and common sensory trigeminal nuclei in python. J. Comp. Neurol., 189, 555 – 572 (1980).

NACHLAS, M. M., TSOU, K. C., SOUZA, E., CHENG, C. S., and SELIGMAN, A. M.: Cytological demonstration of succinic dehydrogenase by a use of a new p-nitrophenyl substituted ditertrazol. J. Histochem. Cytochem., 5, 420 – 436 (1957).

NEWMAN, E. A., GRUBERG, E. R., and HARTLINE, P. H.: The infrared trigeminal-tectal pathway in the rattesnake and in the python. J. Comp. Neurol., 191, 465 – 477 (1980).

NORTHCUTT, R. G., and BUTLER, A. B.: Retinal projections in the northern water snake *Natrix sipedon sipedon* (L.). J. Morph., 142, 117 – 136 (1974).

OTSUKA, N.: Entwicklungsgeschichtliche Untersuchungen am mauthnerschen Zellen von Fischen. Z. Zellforsch., 53, 33 – 50 (1962).

PARENT, A.: Distribution of monoamine-containing nerve terminal in the brain of the turtle, *Chrysemys picta*. J. Comp. Neurol., 148, 153 – 166 (1973).

PALKOVITS, M., and JACOBOWITZ, D. M.: Topographic atlas of catecholamine and acetylcholinesterase-containing neurons in the rat brain. II Hindbrain (mesencephalon, rhombencephalon). J. Comp. neurol., 157, 29 – 42 (1974).

RAMON-MOLINER, E.: Acetylthiocholinesterase distribution in the brain stem of the cat. Ergebn. Anat. Entwickl.-Gesch., 46, 1 – 53 (1972).

REPÉRANT, J., and RIO, J.-P.: Retinal projections in *Vipera aspis*. A reinvestigation using light radioautographic and electron microscopic degeneration techniques. Brain Research, 107, 603 – 609 (1976).

SCHROEDER, D. M.: Retinal afferents and efferents of an infrared sensitive snake, *Crotalus viridis*. J. Morph., 170, 29 – 42 (1981a).

SCHROEDER, D. M.: Tectal projection of an infrared sensitive snakes, *Crotalus viridis*. J. Comp. Neurol., 195, 477 – 500 (1981b).

SCHROEDER, D. M., and LOOP, M. S.: Trigeminal projections in snakes possessing infrared sensitivity. J. Comp. Neurol., 169, 1 – 13 (1976).

SENN, D. G., and NORTHCUTT, R. G.: The forebrain and midbrain of some squamata and their bearing on the origin of snakes. J. Morph., 140, 135 – 152 (1973).

SETHI, J. S., and TEWARI, H. B.: Histoenzymological mapping of acetylcholinesterase and butyrylcholinesterase in the diencephalon and mesencephalon of *Uromastix hardwickii*. J. Hirnforsch., 17, 335 – 349 (1976).

SHEN, S., GREENFIELD, P., and BOELL, E. J.: The distribution of cholinesterase in the frog brain. J. Comp. Neurol., 102, 717 – 743 (1955).

STANFORD, L. R., SCHROEDER, D. M., and HARTLINE, P.: The ascending projection of the nucleus of the lateral descending trigeminal tract: An nucleus in the infrared system of the rattlesnake, *Crotalus viridis*. J. Comp. Neurol., 201, 161 – 173 (1981).

TABER, E., BRODAL, A., and WALBERG, F.: The raphe nuclei of the brain stem in the cat. I. Normal topography and cytoarchitecture and general discussion. J. Comp. Neurol., 114, 161 – 187 (1960).

TORVIK, A.: Afferent connections to the sensory trigeminal nuclei, the nucleus of the solitary tract and adjacent structures. J. Comp. Neurol. 106, 51 – 141 (1956).

TUGE, H.: Somatic motor mechanisms in the midbrain and medulla oblongata of *Chrysemys elegans* (Wied). J. Comp. Neurol., 55, 185 – 271 (1932).

UEDA, S., TAKEUCHI, Y., and SANO, Y.: Immunohistochemical demonstration of serotonin neurons in the central nervous system of the turtle (*Clemmys japonica*). Anat. Embryol., 168, 1 – 19 (1983).

ULINSKI, P. S.: Tectal efferents in the banded water snake, *Natrix sipedon*. J. Comp. Neurol., 173, 251 – 274 (1977).

VILLANI, L., CIANI, F., and CONTESTABILE, A.: Ultrastructural pattern of acetylcholinesterase distribution in the cerebellar cortex of the quail. Anat. Embryol., 152, 29 – 41 (1977).

VILLANI, L., CIANI, F., and CONTESTABILE, A.: Electron microscope histochemistry of acetylcholinesterase distribution in the optic tectum of teleosts. J. Hirnforsch., 20, 539 – 551 (1979).

VILLANI, L., and CONTESTABILE, A.: Cytochemical study of cholinesterase in the normal and retino-deprived optic tectum of reptiles. J. Hirnforsch., 23, 55 – 66 (1982).

WÄCHTLER, K.: The distribution of acetylcholinesterase in the cyclostome brain. I. *Lampetra planeri* (L.). Cell Tiss. Res., **152**, 259—270 (1974).

WÄCHTLER, K.: The distribution of acetylcholinesterase in the cyclostome brain. II. *Myxine glutinosa*. Cell Tiss. Res., **159**, 109—120 (1975).

WAWRZYNIASK, M.: Chemoarchitektonische Studien am Tectum opticum von Telesotiern unter normalen und experimentellen Bedingungen. Z. Zellforsch., **58**, 234—264 (1962).

WILSON, P. M., and MATSON, C.: Acetylcholinesterase staining in an interpedunculotegmental pathway in four spe-cies: *Procavia capensis* (dassie), *Cavia poecllus* (Guinea-pig), *Trichosurus vulpecula* (brush-tail possum) and *Rattus rattus* (hooded rat). Brain Research, **201**, 418—422 (1980).

Address:

Prof. T. KUSUNOKI
Dept. of Anatomy
Yokohama City University
School of Medicine
3—9, Fukuura,
Kanazawa-Ku, Yokohama,
236, Japan

INDEX